図解捕縄術

An Illustrated Guide to Samurai Bondage

Book One

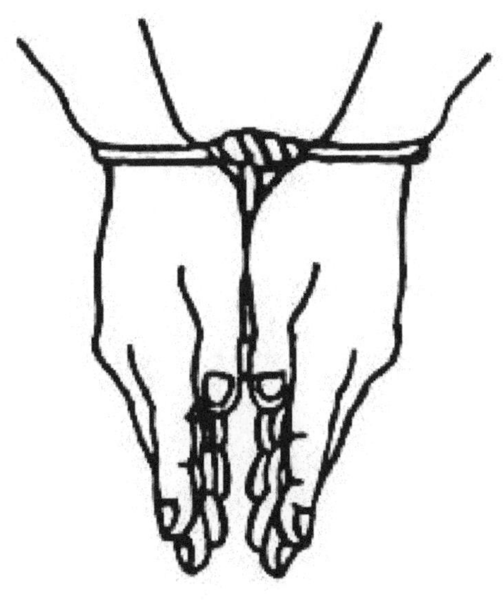

藤田西湖著

By FUJITA SEIKO

シャハン・エリック訳

Translated by Eric Shahan

ISBN: 978-1-950959-11-2

Table of Contents	Page

"Okabiki" Police Constables and a "Zainin" Criminal 岡引と罪人
Photo by Renjo Shimooka 下岡 蓮杖 （1823 -1914）

Translator's Introduction

In Japanese the art of bondage is called Hojo Jutsu. It is the art of restraining a person with rope for a specific purpose, for example arresting, transporting, presenting a prisoner or in preparation for an execution. In some cases bondage was used as a punishment unto itself as in Sarasu, when a criminal is humiliated by tying them up and placing them on the side of the road for one day or several days.

My initial plan was to translate this book as a single volume. However, considering the volume and difficulty of the text along with a desire to simplify the layout, I decided to extend the translation over two volumes. Presenting the elaborate ties as clearly as possible was more important than cramming it all into a single large volume.

Illustration of punishment by Sarasu, humiliation. These priests have broken their Buddhist vows and must spend between 1 ~ 3 days in this fashion. From *Illustrated Guide to the Tokugawa Bakufu Criminal Justice System* by Fujita Shintaro 1893.

Descriptions of Techniques

The man that compiled this book, Fujita Seiko (1899 – 1966,) found a great number of old manuals, called Densho, and reproduced them. The descriptions that accompany the illustrations can vary from extremely detailed to extremely vague. Unfortunately, some of these documents contain little more than titles and illustrations since the details of the techniques would be taught orally. Only members of the schools the documents originate from would know exactly how to tie the knots and their true purpose. However, thanks to Fujita Seiko's efforts we can now appreciate the remarkable breadth of bondage techniques used by Samurai. This book can serve as a tool to understand an art that has nearly gone extinct.

A prisoner is presented to the magistrate in this 19th century photograph. The scene is a re-enactment.

Note on the translation and Kanji readings

Fujita Seiko introduces a number of documents from various Edo Era 1600-1868 schools. Many of these schools are no longer practiced so the meaning and way of reading the names of the techniques is difficult to determine. Unfortunately, only a person who was well versed in these techniques would be able to give the correct reading and meaning for each technique. In addition, following the Meiji Restoration in 1868, many schools closed their doors since they no longer received support from the government. This led to a dramatic decline in the number of practitioners and eventually the techniques taught in these schools were lost.

The readings of the Kanji as well as their English meanings are approximate. Any mistakes regarding those are my own. The purpose of the English translations is to give the reader an idea of the kinds of names that were used and to highlight some recurring themes.

Why are Kanji so hard to read?

Kanji have both a *meaning* and a *reading*. The *meaning* is both what the Kanji stands for on its own as well as how it coordinates with other Kanji. For example, the basic meaning of 木 is "tree," however this meaning can shift when placed next to another Kanji. The Kanji 木 by itself is often read as *Ki* but in the combination 木材 meaning "wood" is read *Moku*-zai, instead of *Ki*-zai. This differs from Chinese, where each character has only one reading. The *reading* can also change depending on if it is alone or joined with another Kanji. If the Kanji is alone, then it may have a common reading. However, if it is joined with another Kanji the way it is read may change. Further, the reading can depend on whether the Kanji appears first or last in the combination.

In addition, some Kanji combinations can have an unpredictable reading. For example the Kanji combination 大人 which is "big + person" meaning "adult." Below is a chart showing some of the possible readings for each of the Kanji.

大	人
Dai	Hito
Tai	Nin
Hiro	Jin
Oo	Bito

So you would expect the combination 大人 to be comprised of one of the readings on the left with one of the readings on the right. However it is read "Otona," for reasons that are not entirely clear even to Japanese people. Starting at least in the 14th century the head village official was called an Otona-bun, written as 咾分 it later shifted to "Otona" written 乙名 and finally the combination 大人 along with the reading of Otona, and the meaning "adult" instead of "head of a village."

Therefore even if you are familiar with all the above readings for the Kanji, it is still possible to "misread" a Kanji combination. Adding to the puzzle is the fact that schools of martial arts used secret readings known only to members of the school.

Illustrated Guide to Samurai Bondage
Book One

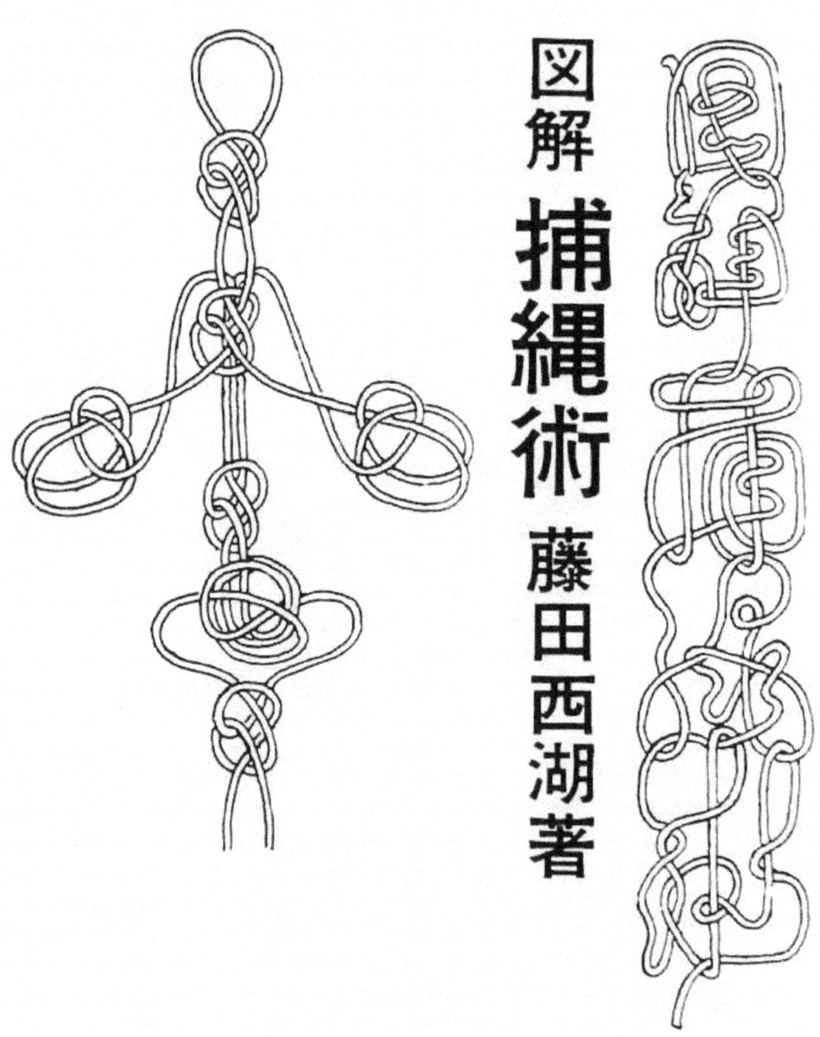

Note: Fujita Seiko drew rope in the shape of the Kanji used in the title of the book.

はしがき

昔　未だ天下が治らず　盛んに戦争が行われていた頃は　武士は戦陣の功名を立てる上にも　敵を生捕りにする業の一として他の武技と共に捕手捕縄の術をも大いに学んだようであるが　世が大平となるに及んで　捕縄術が専ら犯罪人捕縛のための警務用術技として用いられるようになってからは　上位の武士はこれを下役人のする業として漸時縄をも手にしなくなった

徳川時代には　捕方同心以下の者は務めて学んだが　与力以上の武士は学ぶ者も少なかった

明治以降　仮之犯罪人であるにせよ関係なき民間人が乱りにこれを捕へ縄を掛けることは違法であるとされるようになってからは、これを伝え学ぶ者がますます少なくなり　武術家の一部と犯罪人逮捕拘禁護送に当る警務官が　その必要上の捕手早縄と護送用縄の数手を学ぶのみとなった

曽ては、武芸十八般中の一に数えられ伝えられた捕手捕縄術も　今はその術技と共に　その奥儀を伝えた文献資料さえ漸時亡失されつつあるこの時に当り、この術技を学んだ者の一人として　この態を黙視するに忍びず　せめてその術技の一端と故実古伝だけでも残して置きたい念慮から　過去数十年間に亘って蒐集した古伝の文献資料の一部をまとめて　世の研究者のために上梓した次第である

昭和三十九年　秋

藤田西湖

Illustrated Guide to Samurai Bondage: Introduction

Long ago, before the world was at peace, wars flourished in every region of our land. Samurai sought to increase their fame and honor by capturing enemies alive and presenting them to their commanders. The Hojo rope techniques, also called Torite, or capturing, were one of the many martial arts studied by the Samurai. However, when the world became peaceful under the control of the Tokugawa Shogunate during the Edo Era of 1600-1868, Hojo rope techniques were reduced to something only for restraining criminals. Hojo rope techniques evolved into a skill solely employed by those who pursued criminals, and higher-level Samurai no longer learned this art. Soon, no one other than lower level Samurai handled rope.

In the Tokugawa Era, the police officers were all Samurai. Their non-Samurai police assistants, known as Dojin, were the ones that actively studied and used Hojo rope techniques. However even among their immediate superiors, the lowest ranked Samurai, known as Yoriki, very few practiced tying. The fact is, those that practiced the Hojo art of tying knots and bindings were few and far between.

Starting in the Meiji Era (1868-1912) it was basically illegal for anyone to practice the Hojo arts who wasn't associated with the capture of criminals. Clearly this caused the number of people studying Hojo binding techniques to further decrease. Torite-Hojo Jutsu is one of the classic 18 Weapons of War, the 18 Martial Arts a Warrior Should Know, an important aspect of Chinese military science which was later imported to Japan. Though the rope arts are part of the catalogue of techniques a warrior should know, they are in danger of being lost to the sands of time.

As possibly the sole practitioner of these arts, allowing this knowledge to drift away was anathema to me. I felt it was imperative I preserved these old techniques and legendary methods. Thus, over the past decade I have set about gathering Densho, or documents containing the teachings of a school's martial arts. I have collected these rare documents in this volume to make them available to researchers.

Written in the fall of Showa 39 1964
Fujita Seiko

Translator's Note:
Fujita Seiko wrote a whole book on the *Bugei Juhappan* or 18 Weapons of War. In it he states,

The first person to use the term the *Eighteen Weapons of War* was a man who lived in Ming Dynasty China by the name of Sha Cho Sei 謝肇淛 (1567-1624). In the fifth volume of Gozasso 五雑組 or *Five Scrapbooks of Detritus* under the section "Management of People" we can find a description of the Eighteen Weapons:

1. 弓 Yumi - Bow
2. 弩 Do/Oyumi - Crossbow
3. 鎗 Yari - Spear
4. 刀 Katana - Sword
5. 劍 Tsurugi - Double Edged Sword
6. 矛 Hoko/Hoko Yari - A halberd-spear
7. 盾 Tate - Shield
8. 斧 Ono - Axe
9. 鉞 Ono - Axe (Larger and possibly double bladed)
10. 戟 Geki/Hoko - Halberd
11. 鞭 Muchi - Whip
12. 鐗 Kan - Metal Truncheon
13. 撾 Utsu/Kan - Pole-Pick
14. 殳 Fu - Staff with spike on the end
15. 叉 Sasumata - Pole with trident or fork at the end
16. 耙 Ha - Rake
17. 綿縄套索 Wataju Tosaku - Restraining with Rope
18. 白打 Hakuda - Bare handed techniques

例　言

図解捕縄術は　武術としての捕縛用縄法を主として書いたもので　刑罰用処刑縄まで書くのを目的としていない　従って　処刑用首斬縄　晒縄は　他の縄法にも仕用されている関係上載せたが　磔縄　火焙縄　拷問用各種責縄　試劒用各種仕様縄等の類は　本書には一切省略した

又　棒縛　梯子縛　柱縛等他の器具道具を用ゆる縛術法も大小刀縛の外は凡て　これをはぶいた

これらの縄法はいずれ折を見て　拷問用各種責縄　刑罰用処刑縄　試劒用仕様縄　器具道具仕用の各種縛り方　其他変形変態の縛り方一切はまとめて別冊として上梓する心算である

Foreword

This book, *An Illustrated Guide to Samurai Bondage,* introduces the numerous bondage and restraint techniques used in the classical martial arts of Japan. Since the rope and knot techniques related to interrogation and execution of criminals are outside the parameters of this book they have not been included. However, when it is part of the overall curriculum of a martial arts school, I have included some bondage techniques related to preparing prisoners for execution, beheading and Sarasu, or public exposure and humiliation. The bondage used to prepare prisoners for crucifixion, burning at the stake and interrogation along with knots used to tie prisoners, both living and dead, for test cutting with swords will not be included.

In addition, instructions for tying knots to stakes, ladders and columns as well as knots and ties associated with the long and short sword will be eliminated. It is my intention to publish, at a later date, rope and knot techniques related to torture and discipline, official punishments, executions and test cutting. The text cutting will cover how to restrain both living and dead prisoners sentenced to be used for test cutting swords. I hope to include information on other tools and techniques related to Hojo along with all the many variations and permutations of the techniques.

Fujita Seiko

捕縄術とは

捕縄術とは、人を捕え縄を以て縛る術で、一名、取縄術、捕縛術、縲繋之術、伽術、俰術ともいった。支那ではこれを綿縄套索といい、略してただ索ともいった。

この術の初めは、戦場において、敵の捕虜を拘縛したり、乱棒狼藉を働く者を制禦拘禁するために行われたものであるが、後には、専ら犯罪者の捕縛拘禁脱走逃亡を防ぐために用いられるようになったのである。

縄掛方法も、初めは只単に敵の自由を拘束し、乱棒狼藉脱走逃亡の出来ぬような縛り方だけであったのが、後には、大将には大将縄、士卒には士卒縄、下郎には下郎縄と一見その身分階級の見別がつくような縄掛方法を用いたのがいつしか定法となり、ついにはその身分階級ばかりでなく、職柄によっても縛り方が定められ、武家には武家、庶民には庶民、僧侶には僧侶、神官には神官、山伏、行者、婦人、子供、盲目、非人等々、みなそれぞれ一定の縄掛方式によって施縄方法を違えると共に、犯罪の事項、

Chapter 1
What is Hojo Jutsu?

Hojo Jutsu is the martial art that describes how to use rope and knots to bind people. Other names for this include:

取縄術	*Tori Nawa Jutsu* – The art of arresting with rope
捕縛術	*Hobaku Jutsu* – The art of seizing and binding
緊縛之術	*Kinbaku no Jutsu* – The art of binding and restraining
伽術	*Togi Jutsu* – Attendant techniques
俰術	*Yawara Jutsu* – Soft technique (same as Jujutsu)

In China it is called 綿繩套索 *Wataju Tosaku*, Restraining with Rope. It is often abbreviated to the last Kanji only 索 *Saku.*

These techniques began on the battlefield to restrain prisoners taken in battle and to keep those that resist violently under control. Later, the techniques became exclusively for restraining prisoners and preventing their escape.

Initially, when wrapping the rope and tying the knots, the objective was to simply rob the prisoner of the ability to effectively resist or escape. However, later, if a Taisho, or general, were captured, he would be tied up in a Taisho specific manner. If it was a soldier, then he would be tied with a soldier tie and a servant with a servant tie. The technique evolved so a person could tell at a glance the relative status of a prisoner by the way they were tied up. Eventually, the methods were refined beyond simply the class of person (Samurai, Merchant, Craftsmen, Farmer) to a system that had different rope and knot methods of each profession.

Samurai ties were used on Samurai, peasant ties were used on peasants, Buddhist monk ties were used on monks and shrine priests ties were used on Shinto shrine priests. The ways of tying up people depended on their status and differed if the criminal was a mountain ascetic, travelling monk, married woman, child, blind person, a Hinin (non-human caste) and so on. Further, the way the rope was tied would differ from crime to crime. Seeing how a prisoner was bound would indicate the severity of the crime, whether minor or serious. Thus if the knot appropriate for the crime and for the status of the person were tied, there would be no problem.

罪の軽重等まで一見判別できるような掛け方をするのを定法とするようになった。従って定法通りの縄をかけるのはよいが、定法違いの縄を掛けるのは、掛ける者の落度、掛けられる者の恥辱ともなった。

徳川時代一般に用いられた縛縄法としては、

侍には　二重菱縄
庶民（雑人）には　十文字、割菱、違菱、上縄
出家（僧侶）には　返し縄、鷹羽
社人（神官）には　注連縄、鳥居懸
山伏には　笈摺縄
婦人には　乳掛縄
子供には　稚児縄
座頭には　座頭縄
対決には　羽附縄
剛力者には　足固縄
縄抜けの巧みな者には　留縄
罪人追放請渡には　介縄、贈縄、渡し縄
晒物には　晒縄
斬罪には　首切縄、切縄、斬縄、剪縄、落花、捨縄、伏縄
火付には　火付縄
非人には　非人縄

等々をかけるのを定法とした。もっとも流派によってその名称、縄縛法に相違はあった

However, if the wrong type of tie or restraining knot were used, the person who tied the knot would be mortified and the person being restrained would be insulted.

The most commonly used Hakujo Ho, Restraining with Rope Techniques, in the Tokugawa Era (1600-1868) were:

- Samurai should be tied with a Double Water Chestnut Tie
- Peasants should be restrained with a Cross Shaped Tie, Split Water Chestnut Tie, Differing Water Chestnut Tie or Upper Tie
- Travelling Buddhist monks should be restrained with Returning Tie or Hawk Wing Tie
- Those working at a Shinto shrine should be restrained with Churen Tie or Torii (Shinto gate) Tie
- Mountain ascetics should be restrained with Oizuiri Tie (Oizuiri is the traditional white kimono-shirt worn on pilgrimages)
- Married women should be restrained with Over the Breast Ties
- Children should be restrained with Children Ties
- Blind masseuses should be restrained with Blind Masseuse Ties
- Judges should be restrained with Fastened Wing Ties
- Men of great strength should be restrained with Leg Lock Ties
- Tricksters adept at slipping free of their bonds should be secured with Stopping Ties
- When transferring or moving a prisoner, Arresting Tie, Gift Tie or Transfer Tie should be used
- Those set for Sarasu, or being set out by the side of the road as a humiliating punishment, are restrained with Sasrasu Tie
- Those sentenced to be beheaded are restrained with Decapitation Tie, Cut Tie, Cut Tie 2, Chop Tie, Fallen Flower Tie, Pick-up Tie (as in to pick up the body after execution)
- Arsonists are secured with Arsonist Tie
- Hinin or Non-human should be secured with Non-human Tie

In other words there are as many kinds of ties as there are kinds of people. There are of course, differences in the names and ways of tying depending on the Ryuha, or martial arts school.

Note: The final Kanji in the names of the Knots is *Nawa* 縄 which means *rope*, but I felt the word tie fit the overall concept better.

Hojo Densho

捕縄伝書

Historical Samurai Bondage Documents

Translator's Note:

The following section presents transcripts of a type of martial arts document called Mokuroku. Mokuroku are a catalogue of techniques from a school of martial arts. These contain very little, or sometimes no instructional information but can be considered a "rank" in schools of martial arts. Instead of a "black belt," a student receives Mokuroku indicating a certain level of proficiency. The criteria and type of Mokuroku varied widely from school to school but was a common feature. Fujita Seiko has transcribed these rare documents from calligraphy to typed characters.

Some of the documents are extremely vague and do not use Kanji to write the words. Since there are many words with the same reading but different meanings in Japanese, it is difficult to determine the meaning for some sections. Entries with a (?) beside the name indicate that the meaning could not be determined.

The word Musubi occurs frequently in this book. Generally I have translated Musubi as *Knot* when it refers to just a knot and *Tie* when it refers to a binding process made up of multiple knots. Sometimes the word Shibari is used, which can also mean "to tie up," but I have translated this as *Restrain* to delineate the two words.

Though the information in these Mokuroku is quite spare, Fujita Seiko introduces more detailed information about some of these techniques later in this book.

各流で行なわれた捕縄術

荒木流

早縄　一寸縄　有髪縄　鋏縄　鑰縄　夜縄　以上五筋

堅縄　常之縄　位之縄　釣之縄　論之縄　切縄　以上五筋

御家流

本縄　片手縄　追置縄　襟縄　渡し縄　防子縄　羽織下　以上七筋

気楽流　本流派は戸田流より出　戸田流も同

早縄　本縄　盗賊縄　女縄　牢入縄　首切縄　坊頭縄　社人縄　士縄

山伏縄　渡縄　羽付縄　馬上羽付　御前縄　追放縄

荒木流 **Araki Ryu** Techniques from the Araki School		
荒木流 早縄 一寸縄 有髪縄 鈌縄 錀縄 夜縄 以上五筋 堅縄 常之縄 位之縄 釣之縄 論之縄 切縄 以上五筋	Hard Tie Knot, 5 Techniques:	Fast Tie Knots, 5 Techniques:
	Normal Tie	3cm Tie
	Position Tie	Hair Tie
	Hook Tie	Steel Tie
	Argument Tie	Key Tie
	Cut Tie	Night Tie

<table>
<tr><th colspan="2">御家流 Oie Ryu
Seven Techniques From the Oie School</th></tr>
<tr><td rowspan="7">御家流
本縄
片手縄
追置縄
襟縄
渡し縄
防子縄
羽織下
以上七筋</td><td>Main Tie</td></tr>
<tr><td>One-Handed Tie</td></tr>
<tr><td>Follow Through Tie</td></tr>
<tr><td>Collar Tie</td></tr>
<tr><td>Passing Over Tie</td></tr>
<tr><td>Child Guarding Tie</td></tr>
<tr><td>Under the Haori Coat Tie</td></tr>
</table>

気楽流 Kiraku Ryu
Techniques from the Kiraku School
Note by Fujita Seiko: This school evolved out of Toda Ryu. The techniques in Toda Ryu are the same.

気楽流 本流派は戸田流より出 戸田流も同 早縄 本縄 盗賊縄 女縄 牢入縄 首切縄 坊頭縄 社人縄 士縄 山伏縄 渡縄 羽付縄 馬上羽付 御前縄 追放縄	Passing Over Tie	Fast Tie
	Fixed Wing Tie	Main Tie
	Mounted on a Horse Tie	Thief Tie
	Noble Person Tie	Women's Tie
	Banishment Tie	Prison Cell Tie
		Beheading Tie
		Youth Tie
		Shrine Priest Tie
		Samurai Tie

志真古流

柳縄の由来ト申ハ天地開闢ヨリ天神第七代伊弉諾　伊弉冊尊ニテ渡ラセ玉フ其時　天照大神顕シ玉フ其御代ヨリ子安ノ縄初リタリ　天神七代地神五代以来血脈六筋ノ縄出来タリ　天ヨリ不動鎖傳ノ縄渡レリ　一筋ノ縄ニ七筋宛子縄ヲ附親縄共ニ四十八筋ト申縄ニハ九曜七曜ヲ奉表候流ハ万流師者ハ近ト申セ共我等渡ト申ハ越前國櫻埸釆女正ヨリ以来血判令傳授者也

山伏縄　坊主縄　羽飼〆縄　武者縄　国越縄　対決縄　争ノ縄　指出シ縄　晒シ縄
払　縄　掛解縄　渡シ縄　追放縄　百姓縄　女早縄　乳隠縄　手房留縄　てんほこ縄
夜ノ縄　番不入縄　重罪人晒縄　縄手錠　大巻縄　居貴縄　釣　縄　花曼縄　海老縄
落花縄

志真古流 Shima Ko Ryu
Techniques from the Shima Ko School

Translator's Note: The beginning of this Shima Ko School document contains an origin story for the school. It sounds very mysterious and includes a lot of references to Japanese deities. The translation incudes information about the deities that are not directly stated but added for clarity.

The Yanagi "Willow" Rope Techniques, originated with the Tenchi Kai Byaku 天地開闢 the time when all of Japan was created, first recorded in the *Kojiki* in 712 AD. The Seven Generations of Kami emerged one after the other from the newly created earth. When the techniques were going to be passed from Izanagai to mankind, Amaterasu Omikami, the goddess of the sun and the universe, appeared and delivered the first rope. The lineage of six rope techniques comes to us directly from the bloodline of the 5th gods to emerge Otonoji and Otonobe as well as the 7th to emerge Zanagi and Izanami.

From heaven we also received a set of seven rope techniques from the Immovable Chain (possibly referring to Fudo Lord of Light/Fudo Myo-o. See the end of this volume for an introduction to Lord Fudo.) When added together the original group of techniques inherited the total is 48 Hands.

The rope represents the Seven Celestial Bodies in the Sky. The sun, moon, Mars, Mercury, Jupiter, Venus and Saturn. It also represents the Nine Bodies in the Sky, which includes the Seven Celestial Bodies in the Sky in addition to the two mythical bodies Rahu (solar eclipse) and Keta (lunar eclipse.) The heads of all the schools of martial arts that teach rope techniques are said to be similar. Everyone in our school receives transmission through blood vow since Lord Sakuraba Shujosei of Echizen Domain.

Translator's Note: *The 48 Techniques of Sumo* is a famous example of the use of the number 48.

<table>
<tr><th colspan="4">Techniques from the Shima Ko School</th></tr>
<tr><td rowspan="10">山伏縄 坊主縄 羽飼メ縄 武者縄 国越縄 対決縄 争ノ縄 指出シ縄 晒シ縄
払縄 掛解縄 渡シ縄 追放縄 百姓縄 女早縄 乳隠縄 手房留縄 てんほこ縄
夜ノ縄 番不入縄 重罪人晒縄 縄手錠 大巻縄 居責縄 釣縄 花髪縄 海老縄
落花縄</td><td>Night Tie</td><td>Buddhist Priest Tie</td><td>Mountain Ascetic Tie</td></tr>
<tr><td>Unguarded Tie (No need for a guard tie)</td><td>Quick Release Tie</td><td>Wealthy Young Man Tie</td></tr>
<tr><td>Exposure for a Severe Crime Tie</td><td>Transfer Tie</td><td>Wings Pinned Back Tie</td></tr>
<tr><td>Rope Handcuffs</td><td>Banishment Tie</td><td>Samurai Tie</td></tr>
<tr><td>Full Body Wrap</td><td>Peasant Tie</td><td>Inter-Country Tie</td></tr>
<tr><td>Securing for Punishment (before being caned)</td><td>Women's Fast Tie</td><td>Judge Tie</td></tr>
<tr><td>Hook Tie</td><td>Concealed Women's Chest Tie</td><td>Struggle Tie</td></tr>
<tr><td>Kaman Tie</td><td>Forearm Lock Tie</td><td>Fingers-free Tie</td></tr>
<tr><td>Shrimp Tie</td><td>Tenhoko (?)Tie</td><td>Exposure Tie</td></tr>
<tr><td>Falling Flower Tie</td><td></td><td></td></tr>
</table>

當慎流 *Toshin Ryu*

Techniques From the Toshin School

The techniques in this school are the same as in the Muso School.

常慎流　当流縄法八夢相流ト同 白洲縄　七曜　乳割　腰縄　腰小手留　渡シ縄　追払縄　大縄　トイ縄 トウ縄　マワシ縄　ツユカケ　早縄トイ　ケサカケ　ケマン　小手カヘシ　芝居縄 番イラズ　ツメ縄　早縄　早縄	Unguarded Tie	Toh (?) Tie	White Sandbar
	Finish Tie	Wrapping Around Tie	Seven Visible Bodies
	Fast Tie #1	Mist Tie	Split Chest
	Fast Tie #2	Toi (?) Fast Tie	Waist Tie
		Kesa Tie (Monk's Robe Tie)	Waist Tie With Wrist Tie
		Keman (?)	Transfer Tie
		Tying Back the Wrists	Banishment Tie
		Presentation Tie	Big Tie
		Toh (?) Tie	Toi (?) Tie

心外無敵流 *Shingai Muteki Ryu*
Techniques from the Shingai Muteki "Invincible" School
Seven Techniques

心外無敵流	
早縄	Fast Tie
四寸縄	12cm Tie
さげをしばり	Sage wo Shibari (Tying off the Sage-o, the rope that is tied to a Katana)
女しばり	Women's Tie
ちごしばり	Chigo Tie (?)
ちや坪しばり	Chiya Hei Tie (?)
棒しばり	Pole Tie
以上七筋	

真之神 ***Shin no Shindo Ryu***

Techniques from the Shin no Shindo School

Shin no Shindo divides our rope techniques into three categories; Fast Tie, Main Tie and Full-Body Tie.

<table>
<tr><td rowspan="6">真之神道流
真之神道流は　早縄　本縄　体中縄の三つに別け
道中縄　腰縄　翅〆　僧縄　女縄　の以上五法とす</td><td>Five Techniques:</td></tr>
<tr><td>Along the Road Tie</td></tr>
<tr><td>Waist Tie</td></tr>
<tr><td>Arm Lock Tie</td></tr>
<tr><td>Monk Tie</td></tr>
<tr><td>Women Tie</td></tr>
</table>

関口新心流

両手ノ早ククリ　羽カヒ緊（手ノ大指結様伝書両ヒジニテ止ル）
番不入（足カカマリ下ニテトメリ）　忍　縄　倒　縄　片手掛
腰　縄（胴ヲ一重マハス）　田　掛（小手チキリ）
三尺縄（是大指ニテ止ル）　以上の外縄法トシテ
早　縄　二つかきくわん　四寸縄　番無縄　切　縄　侍　縄　（以上五）
お家縄　山伏縄　高手縄　羽がい付縄　巻　縄　児女縄　ほたし縄　胸刀之縄
小手縄　また縄　（以上十）
棒しばり　はしごしばり　はしらしばり　はた者しばり　すかきしばり　たゝみしばり
立しばり　俵しばり　二人しばり　片手しばり　（右十）
木馬しばり　からしばり　縄無しばり　二つ声しばり　小脇指しばり　かけきよしばり
てんびんしばり　もつこふしばり　さかしばり　てかね　首かね　（右十）
籠　すまき　筋たち（以上三条極意）

関口新心流 *Sekiguchi Shin Shin Ryu*
Techniques from the Sekiguchi Shin Shin School

Wrap the rope quickly around both hands and pull the wings (arms) back and lock them.
Note by Fujita Seiko:
I have another scroll from the same school that states the opponent's thumbs should be forced into the inside of the elbows of both arms.

When applying the Unguarded Tie be sure the rope wraps around the legs and ties at the bottom. This technique is also known as Shinobi Tie, Fallen Tie and One Hand Tie. When applying Waist Tie make sure the rope goes all the way around the waist. It is also known as Field Lock Tie or Keeping the Wrists Separate Tie.
Sanjaku Nawa : 90 Centimeter Rope
When making 90cm Rope, it shouldn't be thicker than the thumb. That is all regarding this kind of rope.

早縄 Fast Tie
Five Techniques:

Two Ties at the Same Time
12 Cm Rope tie
Unguarded Tie
Cut Tie
Samurai Tie

お家縄 Household Ties
Ten Techniques:

Mountain Ascetic Tie
Wealthy Tie
Bound Wings Tie
Wrapped Tie
Young Girl Tie
Hotashi (?) Tie
Concealed Knife Tie
Wrist Tie
Thigh Tie

棒しばり Tied to a Pole
Ten Techniques:

Tied to a Ladder Restraint
Tied to a Bridge Restraint
Restraining a Signal Flagman
Rough Tie
Folded Tie
Standing Tie
Tied to a Bale of Rice
Two Person Tie
One-Handed Tie

木馬
Wooden Horse Restraints
Ten Techniques:

Empty Restraint
No Rope Restraint
Two Voice Restraint
Fingers Under the Armpits Restraint
Kake Kiyo (?) Restraint
Tenbin (?) Restraint
Motsukofu (?) Restraint
Slope Restraint
Hand Lock
Neck Lock

極意
Inner Secrets
Three Techniques:

Basket
Bare-Handed Wrap
Standing Sinews

直至五傳流

一、早縄之事
最初蛇口の処を右の手首へかけ夫より左の肩より首へまはし左の小手を右の小手にそへ縛る

一、不入番之事
玉二つたて縄足へもかゝる

一、下緒縛之事
両方の手首を袂へ入れ両方の肘へ鞘を通し下緒にてとめること

一、武士縄之事
咽へ紙をあてること

一、本縄之事
玉二つ拵て腕へかけ印付にし縄の両端を合せ菱の如く結び夫より小手を留るなり

一、羽骸付之事
右の手を切付に縛り右の腰へつけ左の手も同様に縛り左の腰へつけ後帯の結目の処にて縄の両端をいぼ結にし帯へ引通すこれは途中など連行時の縄なり

一、番不入之事
本縄の如く縛りて足へかけること

直至五伝流 *Jiki Shi Goden Ryu*
Techniques from the Jiki Shi Goden School

1. Fast Tie
First tie the Snake's Mouth around the captive's right wrist. From there wrap it over his left shoulder and around his neck, down to his left hand. Tie the left hand to the right.

2. Unguarded Tie
Using two coils of rope, secure the captive's arms as well as his legs.

3. How to restrain a captive with the Sage-O of their own sword
Have the Samurai place both hands in his sleeves until the right hand is on the left elbow and the left hand is on the right elbow. Pass the scabbard of his Katana through the open hole in the armpit of his shirt on one side. Pass it through his sleeves and out the other armpit. Finally, tie the arms off with the Sage-O cord.
Translator's Note: This description is very brief, so this is my interpretation.

4. Securing a Samurai with Rope
Paper should be placed around the throat before tying.
Note: This might be to prevent rope burn.

5. Main Tie
Using two coils of rope doubled up, wrap the arms and tie a finishing knot. Bring the two ends of rope together and tie a knot shaped like a water chestnut. Next, tie off the hands.

6. Remains (Bones) of Wings
Loop the rope around the right arm of the captive and then tie it to his right hip. Use the same tie on his left arm and secured to his waist on the left side. Pass the two ends of rope through the back Obi (belt) and tie it off. This tie can also be used when escorting or transporting captives.

7. Guard Unnecessary Tie
This uses Main Tie but incudes a tie around the legs.

<table>
<tr><th colspan="2">夢雙神鳥流 Musoshinto Ryu
Techniques from the Mushoshinto School</th></tr>
<tr><td rowspan="7">夢雙神鳥流
一寸縄　カラナハ　ナカナハ　アラソイ縄　イタマス縄（坊主　女人）　ハマナハ（二尺五
寸）　カキナハ（三尺二寸）（以上七ケ条）</td><td>3cm Tie</td></tr>
<tr><td>Empty Tie</td></tr>
<tr><td>Middle Tie</td></tr>
<tr><td>Rough Tie</td></tr>
<tr><td>Not Painful Tie
For use on monks and women.</td></tr>
<tr><td>Beach Tie uses 75cm rope</td></tr>
<tr><td>Persimmon Tie uses 96cm rope</td></tr>
</table>

藤原流 *Fujiwara Ryu*
Techniques from the Fujiwara School

<table>
<tr><td rowspan="10">藤原流
早之縄 鑰縄 付錄縄 論縄 有髪縄 位之縄 常之縄 真之縄 一寸縄
夜之縄 釣縄 伏縄 切縄</td><td>Night Ties</td><td>Fast Ties</td></tr>
<tr><td>Hook Tie</td><td rowspan="3">Fence Tie

Includes the sub-technique:
Statue Tie</td></tr>
<tr><td>Sprawled Flat Tie</td></tr>
<tr><td>Cut Tie
Executioner’s Tie</td></tr>
<tr><td></td><td>Argument Tie</td></tr>
<tr><td></td><td>Long Haired Tie</td></tr>
<tr><td></td><td>Noble Tie</td></tr>
<tr><td></td><td>Normal Tie</td></tr>
<tr><td></td><td>True Tie</td></tr>
<tr><td></td><td>3cm Tie</td></tr>
</table>

<table>
<tr><th colspan="2">山田新心流 Yamada Shin Shin Ryu
Techniques from the Yamada Shin Shin School</th></tr>
<tr><td rowspan="5">山田新心流
早縄　番不入　本縄　棒縛　短尺　（以上五筋）</td><td>Fast Tie</td></tr>
<tr><td>Unguarded Tie</td></tr>
<tr><td>Main Tie</td></tr>
<tr><td>Stick Tie</td></tr>
<tr><td>Short Tie</td></tr>
</table>

Hojo
捕縄
Rope and Tools for Restraining Prisoners

捕縄

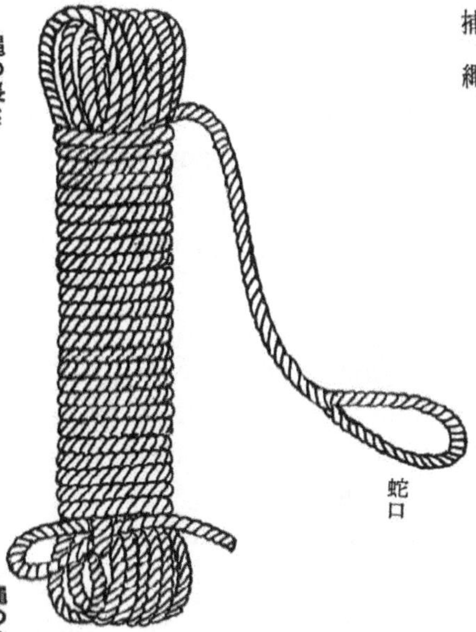

捕縄は、もっとも良質な麻を極めて柔かに打ち、これを三操編の細目にしたものが良いとされている。そしてこれを血で染めたいわゆる血染縄がもっとも良いとされた。血染縄は永年使っても塩気がつかず腐ることがない。また縛るにもすこぶる締りが良い。渋染縄は腐りが早くかつ解け易い。絹縄は丈夫で締りも良いが解け易い欠点もある。江戸時代捕縄は参州宝蔵寺で製した縄がもっとも珍重された。

縄の長さ

縄の長さは流派によって、その縄縛法の繁簡あり、一定しないが、

本縄は三尋半から十一尋
早縄は二尋半から三尋半
鉤縄は早縄の長さと同じか、またそれよりやや短いものが用いられていた。鉤には一個鉤、二個鉤、塀乗鉤等の別があった。

その他三寸縄、五寸縄、七寸縄等がある。

縄の色

縄の色は、古くは四季によって色を違え、またそれに相当する方位に向って縄をかけた。

春は東に向って、青色の縄を用い
夏は南に向って、赤色の縄を用い
秋は西に向って、白色の縄を用い
冬は北に向って、黒色の縄を用い
土用は中央で、黄色の縄を用いたが、後には罪の軽い者には白縄、罪の重い者には青縄、位のあ

Hojo

The best Hojo, or restraining ropes, are made from hemp fibers which have been pounded soft. Ropes braided with three thin strands are said to be best. After it is braided, dying the rope with blood, called Chi Zo-meh in Japanese, is the best. The reason is because it is long-lasting. The salt in the blood will prevent the rope from rotting. In addition, blood-dyed ropes are very easy to tie. Ropes dyed with astringent dyes tend to rot quickly and can come undone. Ropes made of silk are strong and easy to tie, however the knots come untied easily. In the Edo Era the most highly prized rope was made at Sanshu Hozo Temple 参州法蔵寺. Which is in modern day Aichi Prefecture. This temple was founded in the year 701 by the monk Gyōki 行基 (668-749.)

Rope Length

The length of the Nawa, or rope, can vary according to the school, the specific method or the complexity of the knot. Thus, there is no standard length for rope, however I will list some common lengths.

- Rope used for Hon Nawa, Main Tie, is measured in Lengths. One length is the distance from hand to hand when your arms are spread wide. It is from 150~180 cm. Hon Nawa rope is generally from 3.5 to 11 lengths.
- Rope Used for Haya Nawa, or Fast Tie is 2.5~3.5 Lengths.
- Rope used for Kage Nawa or Rope with a Hook, is the same length as Haya Nawa or slightly shorter. There are several different categories of Kage Nawa including ropes with one hook, two hooks and ones with multiple hooks for climbing over walls.

In addition there are ropes such as the 9 cm Rope, 15 cm Rope and 21cm Rope.

Translator's Note: The word Hojo can also be read as Torinawa. Both are correct and can be used interchangeably

Color of Rope

In the past the color of the rope you used was changed according to the season. In addition, it was a common practice to face a certain compass direction when tying a prisoner. The color and season pairings are as follows:

Spring: Face east and use a blue rope
Summer: Face south and use a red rope
Fall: Face west and use a white rope
Winter: Face north and use a black rope
Midsummer: In Doyo, or midsummer, face the center and use a yellow rope.

If the crime is a minor one, then a white rope would be used. A blue rope should be used for serious crimes and if the criminal is a high-ranking person then a purple rope should be used. Lower ranks are tied with a black rope and other ranks should be tied with a rope dyed yellow or light blue depending on their status.

During the reign of the Tokugawa Shogunate (1600-1868) in Edo City, the dyed ropes were known as Yokome, Inspector Rope as well as Marked Rope. The Dojin, low level police that worked at the Machi Bogyo police stations in northern Edo, used a white rope to escort prisoners. Those working at the southern police stations used navy blue rope. Those working at the treasury used a white and black striped rope. The rope used at the Royashiki, or prison, was navy blue.

In the Meiji Era (1868-1912) the practice of dying ropes different colors was abandoned. Instead Hari Nawa, a rope with a metal spike on the end and Kagi Nawa, a rope with a single or double hook on the end, were used. Only a few of the old methods remained, one was Haya Nawa, Rapid tie. There were also two kinds of Hon Nawa, Main Tie, including Goso Nawa, which is used for guarding and transporting prisoners. Further, many aspects of Hojo became standardized. The rope used for transporting and guarding prisoners should be about 7 meters long and about 5 mm in diameter. Standard arrest and restraining rope should be about 5 meters long and about 3.5 mm thick.

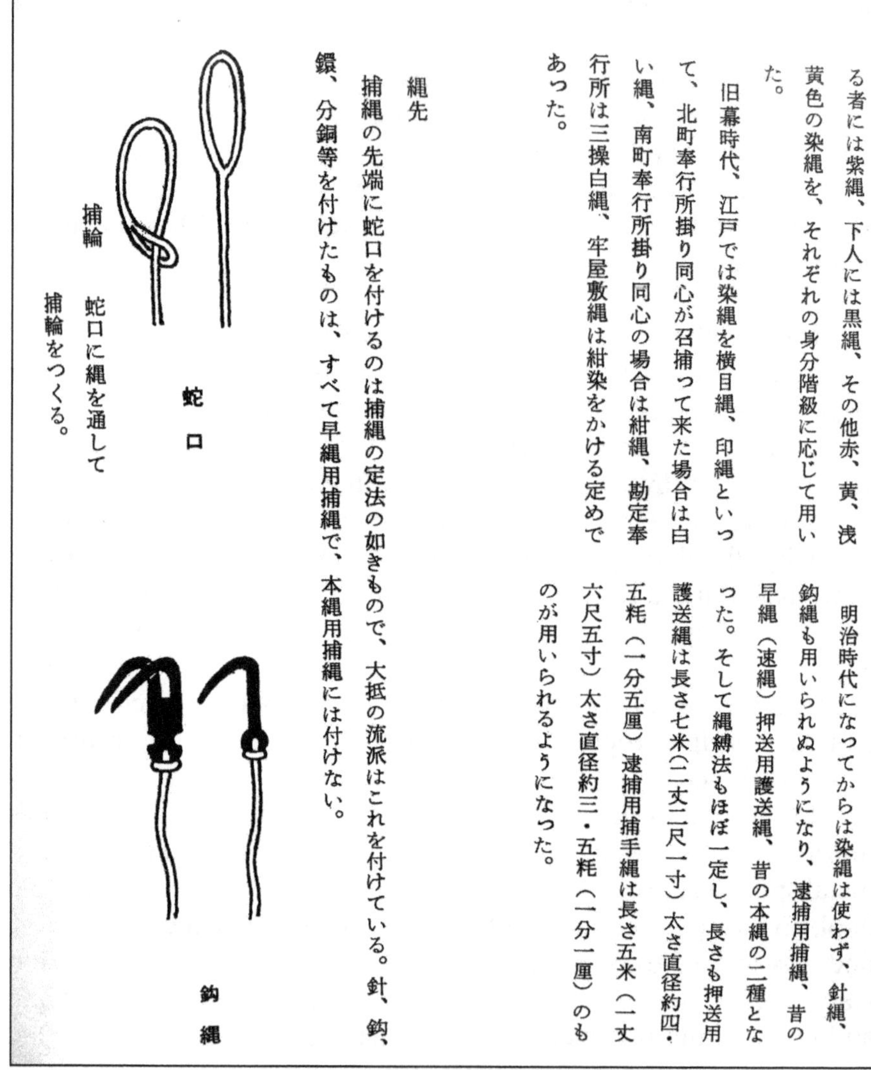

る者には紫縄、下人には黒縄、その他赤、黄、浅黄色の染縄を、それぞれの身分階級に応じて用いた。

旧幕時代、江戸では染縄を横目縄、印縄といって、北町奉行所掛り同心が召捕って来た場合は白い縄、南町奉行所掛り同心の場合は紺縄、勘定奉行所は三操白縄、牢屋敷縄は紺染をかける定めであった。

明治時代になってからは染縄は使わず、針縄、鈎縄も用いられぬようになり、逮捕用捕縄、昔の早縄（速縄）押送用護送縄、昔の本縄の二種となった。そして縄縛法もほぼ一定し、長さも押送用護送縄は長さ七米(二丈二尺一寸)太さ直径約四・五粍（一分五厘）逮捕用捕手縄は長さ五米（一丈六尺五寸）太さ直径約三・五粍（一分一厘）のものが用いられるようになった。

縄先

捕縄の先端に蛇口を付けるのは捕縄の定法の如きもので、大抵の流派はこれを付けている。針、鈎、鐶、分銅等を付けたものは、すべて早縄用捕縄で、本縄用捕縄には付けない。

Nawa Saki
End of the Rope

The Jaguchi, Snake's Mouth, is the standard loop tied to the end of an arresting rope. Most Ryuha (martial art schools) attach this to one end. A spike, hook, ring or weight are only attached to ropes used for Haya Nawa, Fast Tie. They are not used for Hon Nawa, Main Ties.

Jaguchi Snake's Mouth Seizing Loop: A Seizing Loop is made by passing the rope through the Snake's Mouth.	**Kagi Nawa** Rope with metal hook attached

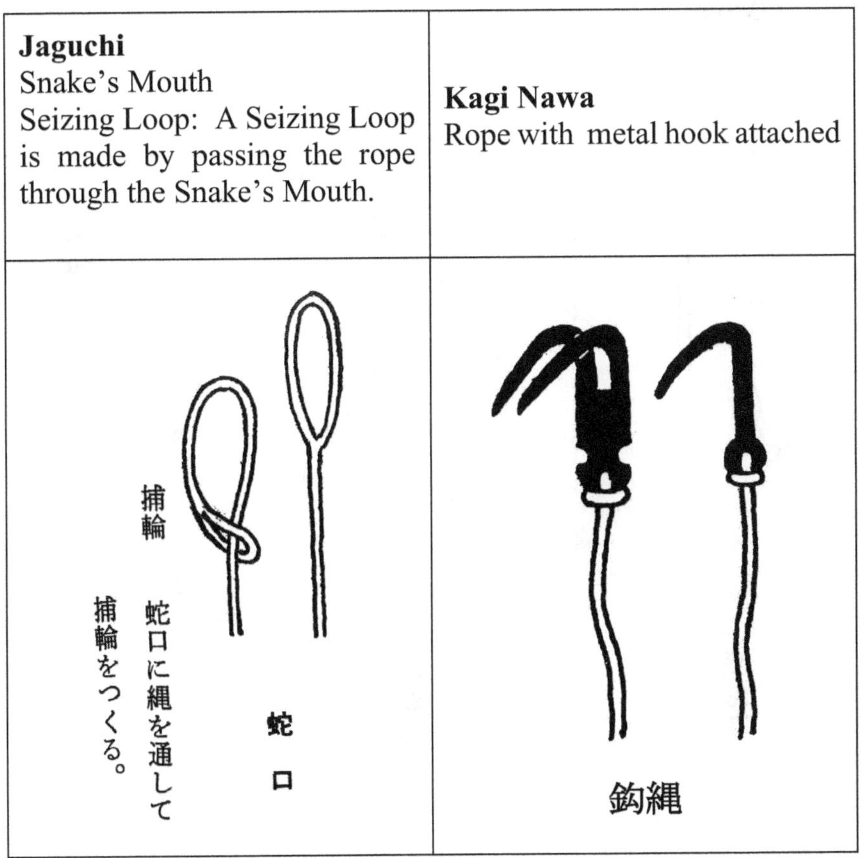

Hojo Rope and Tools for Restraining Prisoners		
Kari Jaguchi Simple Snake's Mouth	**Orikake Nawa** Folded Rope	**Kan Nawa** Rope With Metal Ring Attached The main benefit of the Kan Nawa is how smoothly the rope slides through the ring.
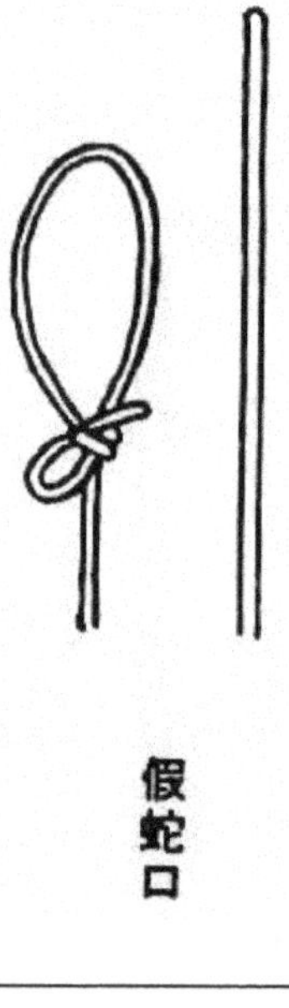 假蛇口	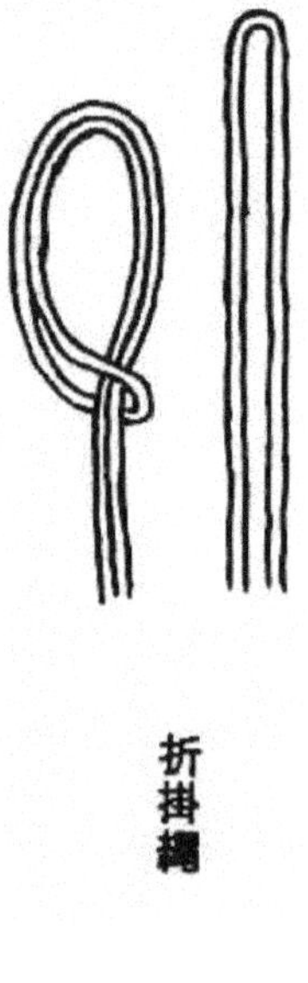 折掛縄	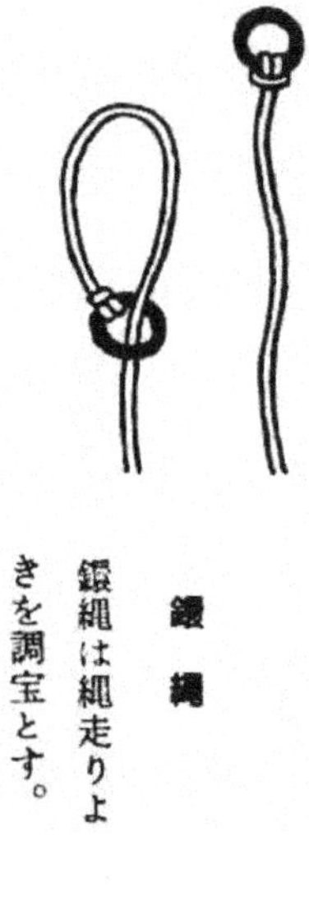 鐶縄 鐶縄は縄走りよきを調宝とす。

Hojo 捕縄 Rope and Tools for Restraining Prisoners		
Fundo Nawa Weighted Rope	**Haya Tejo** Quick Handcuffs	**Hari Nawa** Rope With Metal Spike Attached The Hari Nawa uses the tool Odoshige (used for attaching armor) as a needle to thread the course of the rope when tying.
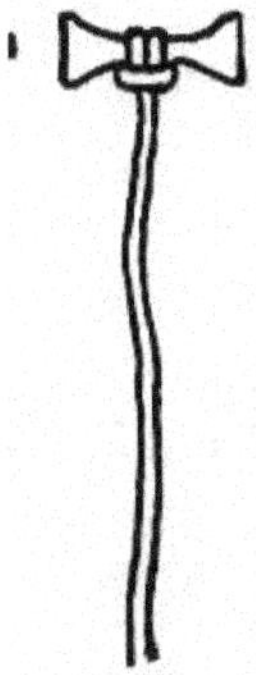 分銅縄	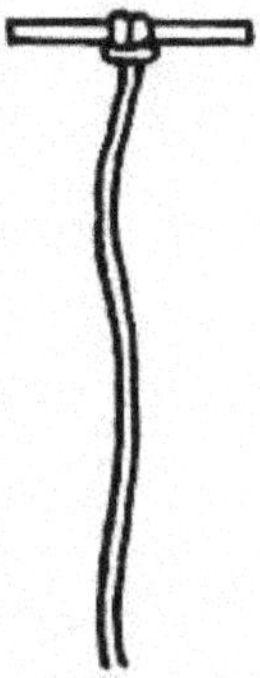 分銅縄 早手錠	 針縄

Hojo Rope and Tools for Restraining Prisoners
捕縄畳様種々 ***Hojo Tatami-Yoh Shu-Shu*** **Examples of Different Ways to Fold the Arresting Rope**
本縄用 ***Hon Nawa Yoh*** **Examples of Coiled Main Tie Rope**
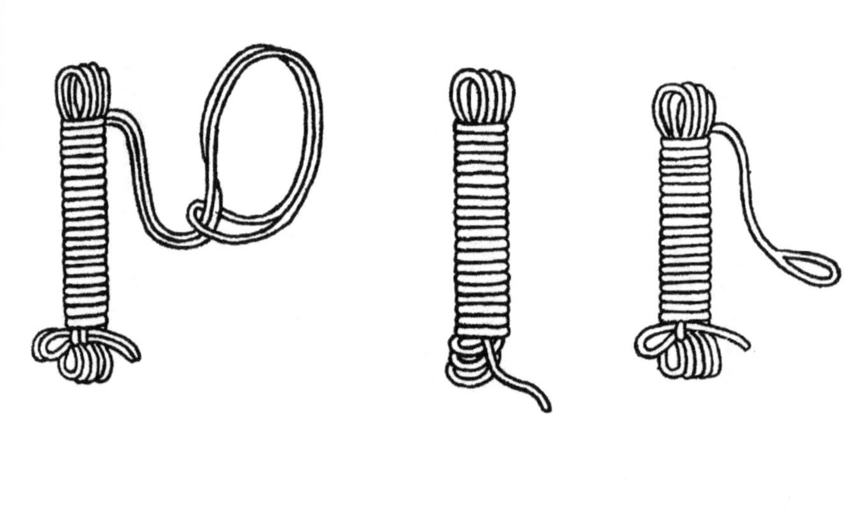

Hojo 捕縄 Rope and Tools for Restraining Prisoners

捕縄畳様種々 ***Hojo Tatami-Yoh Shu-Shu***
Examples of Different Ways to Fold the Arresting Rope

早縄用 ***Haya Nawa Yoh***
Examples of Coiled Fast Tie Rope

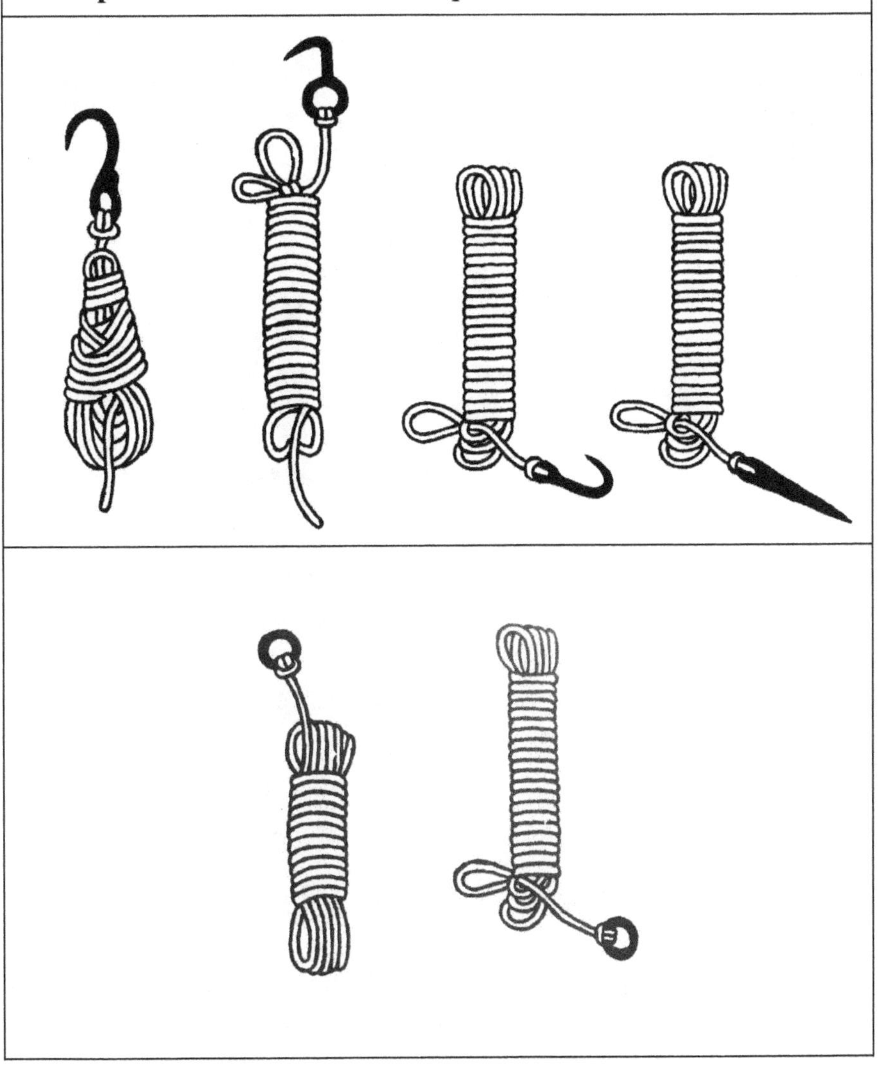

Hojo Rope and Tools for Restraining Prisoners

Coiling the Arresting Rope /Rapid Deploying Rope Two Methods

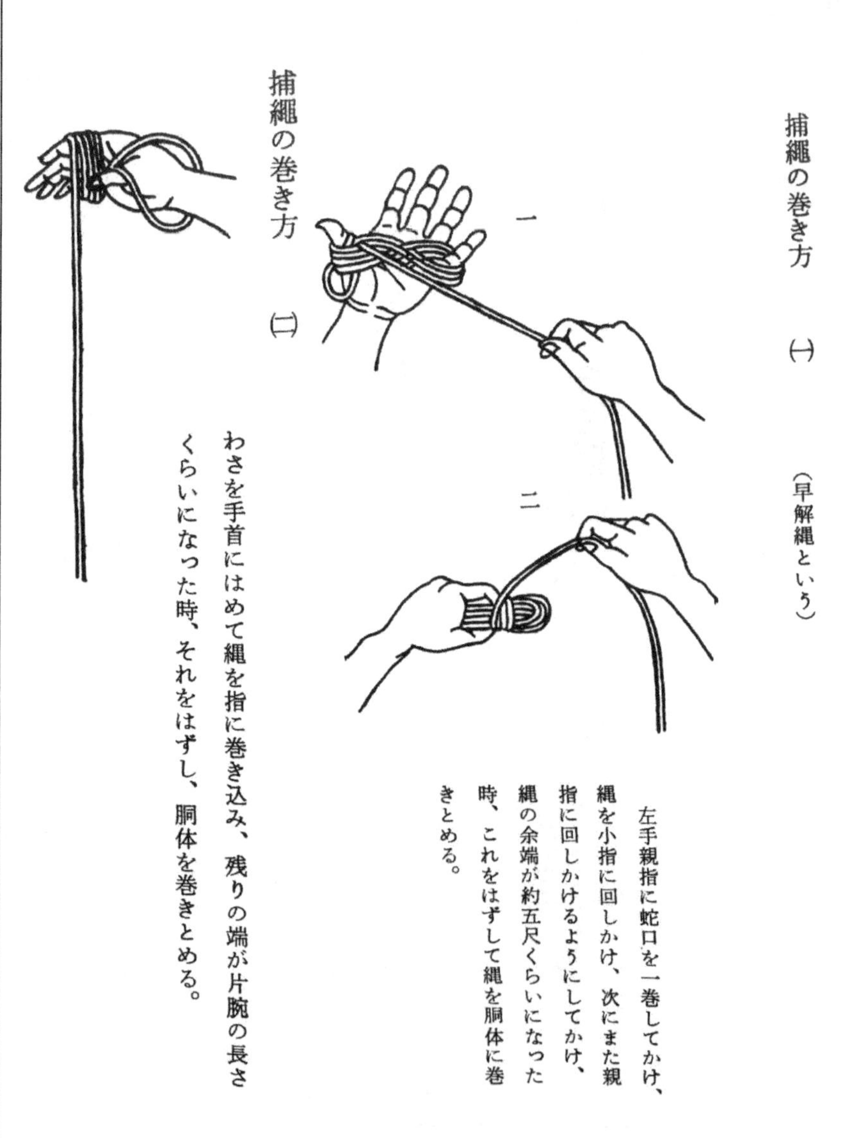

Hojo　Rope and Tools for Restraining Prisoners

Method One: First hook the Snake's Mouth around the thumb of your left hand, then wrap it around your little finger. Continue to wrap the rope back and forth between your thumb and little finger. When there is about 180cm of rope remaining take it off your fingers and begin to wrap the body of the coil.

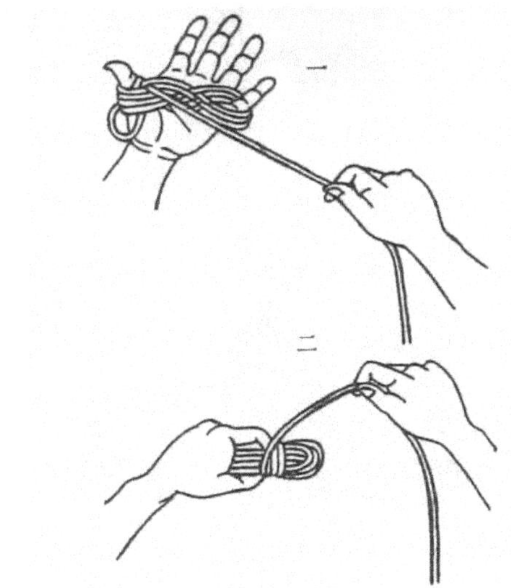

Method Two: Slip the loop over your wrist and begin wrapping the rope around your fingers. When the remaining rope is as long as one arm extended begin to wrap the body of the rope.

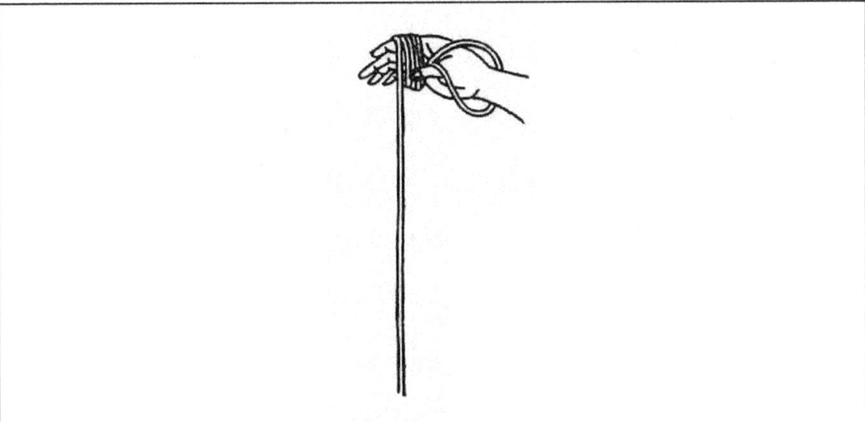

Hojo 捕縄 Rope and Tools for Restraining Prisoners	
Jaguchi: Snake's Mouth and application	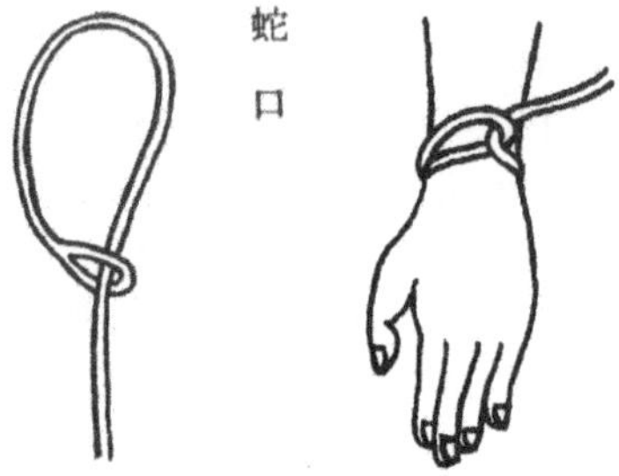 蛇口
Ketsu Jaguchi: Tied Off Snake's Mouth and application	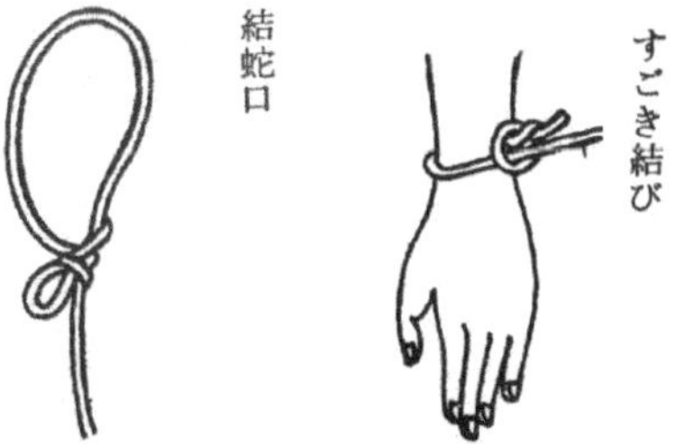 結蛇口 すごき結び
Kamo Sage: **Duck's Wings**	**Sugoki Musubi:** **Seizing Knot**
 かもさげ	 すごき結び

<table>
<tr><th colspan="3">Hojo Rope and Tools for Restraining Prisoners</th></tr>
<tr><td>Go Gyo Musubi:
5 Elements Knot</td><td colspan="2">Kanau Mutsubi:
Prayer Knot</td></tr>
<tr><td>一筋五行結び</td><td colspan="2">一筋叶結び</td></tr>
<tr><td>Orikate Nawa:
Folded Over Rope</td><td colspan="2">Hibari Musubi:
Skylark Tie #1</td></tr>
<tr><td>折掛縄</td><td></td><td>雲雀結び</td></tr>
</table>

Hojo Rope and Tools for Restraining Prisoners

雲雀結び *Hibari Musubi*: Skylark Tie #2

Once you have attached the Hibari Musubi to the fingers you will find the rope does not slip off.

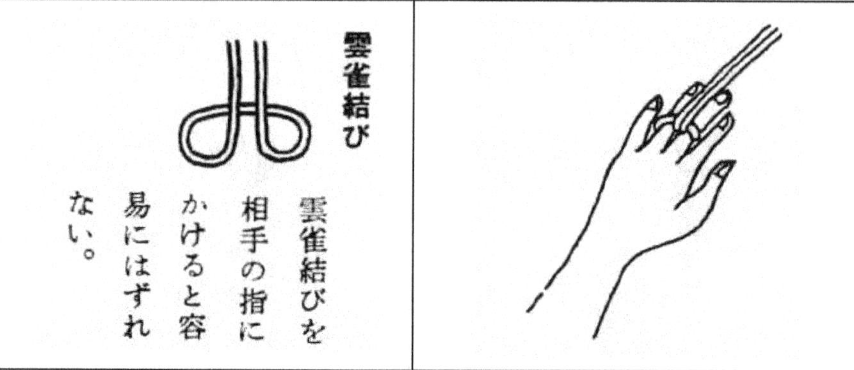

鐶縄 *Kan Nawa*:
Rope With Metal Ring Attached

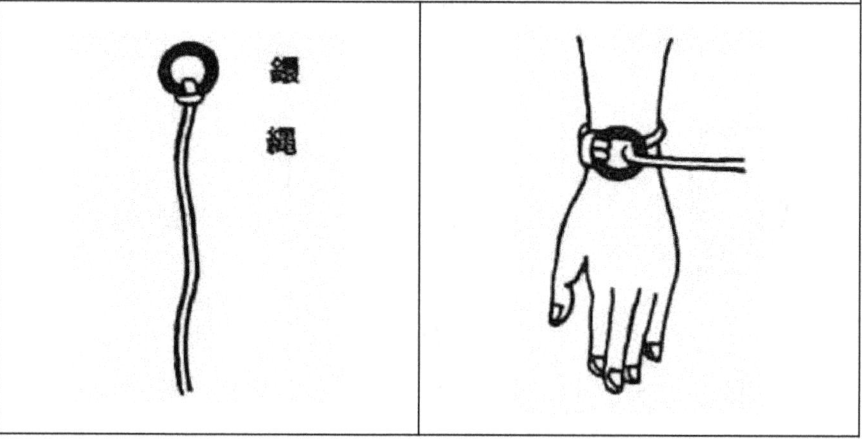

Haya Tejo: Quick Handcuffs

These are two mallet-shaped pieces of hard wood approximately 6 cm long. Attach a piece of rope about 50 ~ 55 cm long. To use, wrap the cord around the wrists just like in Haya Nawa. End with the two pieces between than hands and twist them several times. Finally wedge them in between the hands.

Translator's Note: Fujita called this a Fundo Nawa on page 45.

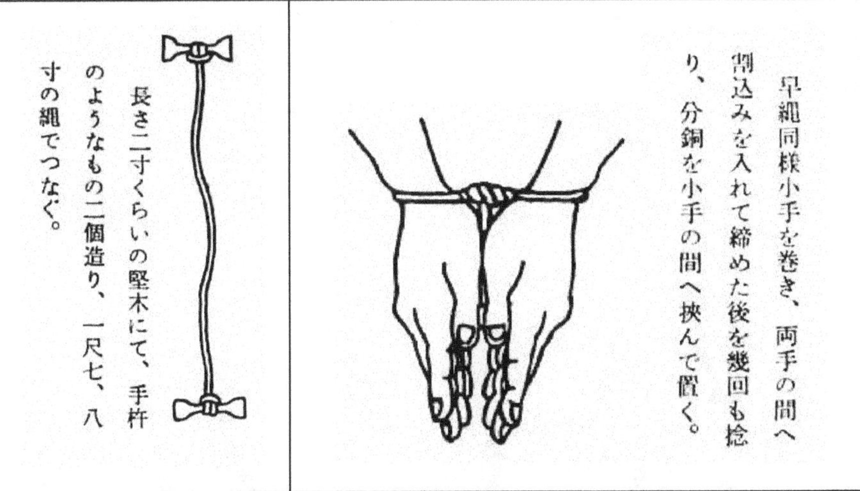

Haya Tejo #2: Quick Handcuffs #2

This is a Standing Pull Device also known as Quick Handcuffs. These stick shaped pieces can be made from either bamboo or brass and are 6 ~ 7 cm long and about 1 cm in diameter. A rope 18 cm long and 1 cm thick is attached. They are used as shown in the illustration.

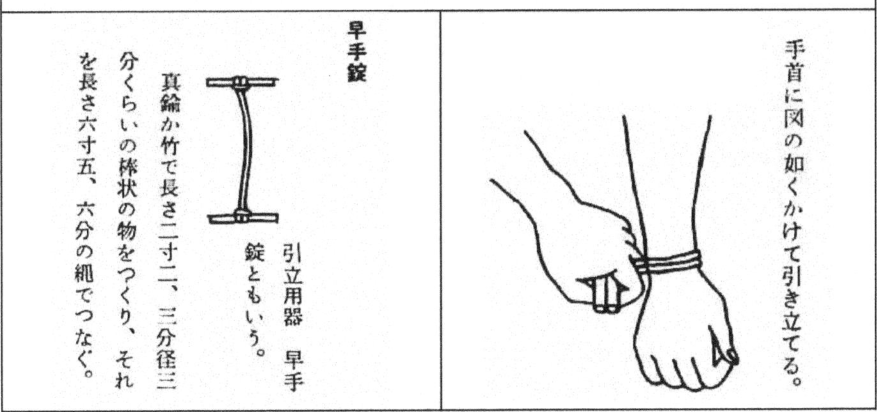

Musubi Kata

結び方

Knots

Musubi Kata 結び方 Knots	
Otoko Mususbi: Men's Knot Other names for this knot include; Oro Musubi, Moro Musubi (Double Knot,) Kaine Musubi (Fence Knot,) Ibo Musubi (Wart Knot,) An Mususbi (Hermiatage Knot,) Shiori (Garden Gate Knot,) Hae-gashira (Fly's Head Knot.)	
How to tie a Men's Knot/ Gate Knot	**How to tie a Men's Knot**
結び方 一 1 二 2 三 3	**男結び** 別名 をろ結び もろ結び 垣根結び 籬結び 疣結び 庵結び 枝折結び 蠅頭 いがら結び ろ結び 結び方 一 1 二 2 三 3

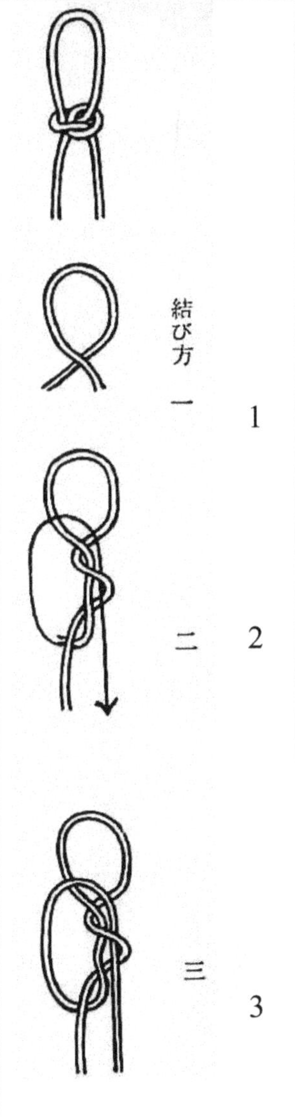

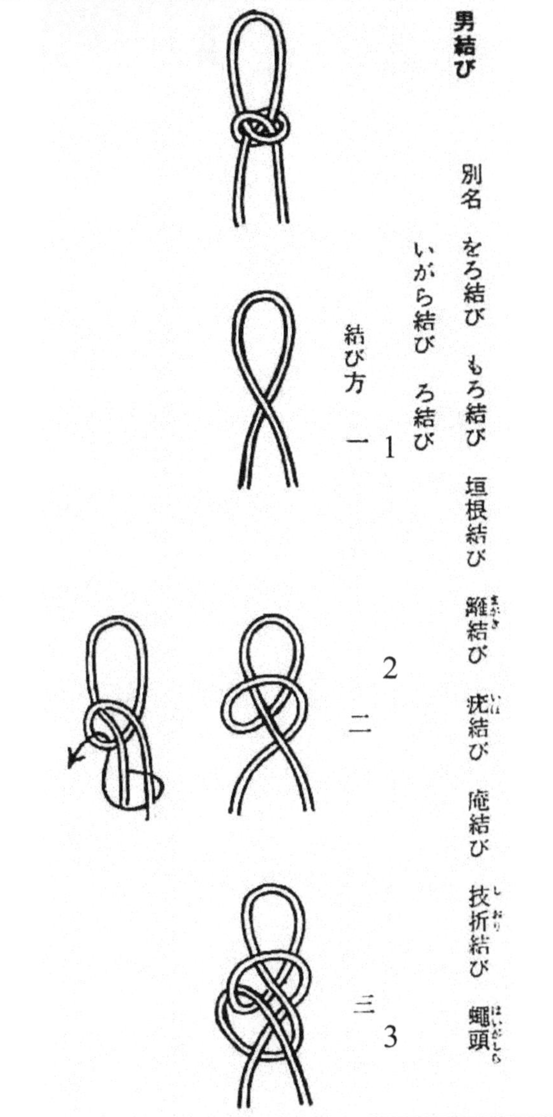

Musubi Kata 結び方 Knots	
Onna Musubi 女結び Women's Knot This is also known as めなご結び Menago Musubi, Girl's Knot	
How to tie Kata Mususbi: Left Half-Tie	**How to tie Onna Musubi: Women's Knot**
左片結び 片結び 結び方 一 1 二 2	結び方 一 1 二 2 三 3

<table>
<tr><th colspan="2">Musubi Kata 結び方 Knots</th></tr>
<tr><th>How to tie Aioi Musubi: Decorative Knot</th><th>How to tie Kata Musubi: Right Half Knot</th></tr>
<tr><td>
相生結び

結び方
一 1

二 2</td><td>
右片結び
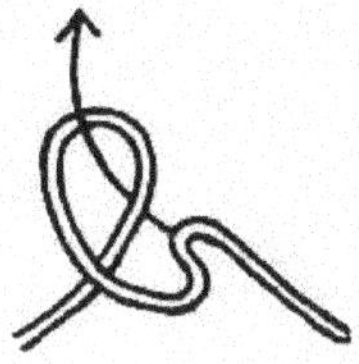
結び方
一 1
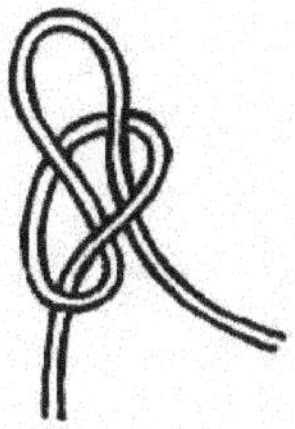
二 2</td></tr>
</table>

Musubi Kata 結び方 Knots

Left: **How to tie Karasu no Kubi: Crow's Neck Knot**
Right: **How to tie Ro Musubi: Mist Knot**
The Mist Knot is also known as the General's Knot. It is tied the same as the Men's Knot. If it is done by crossing to the left, then it becomes the Women's knot.

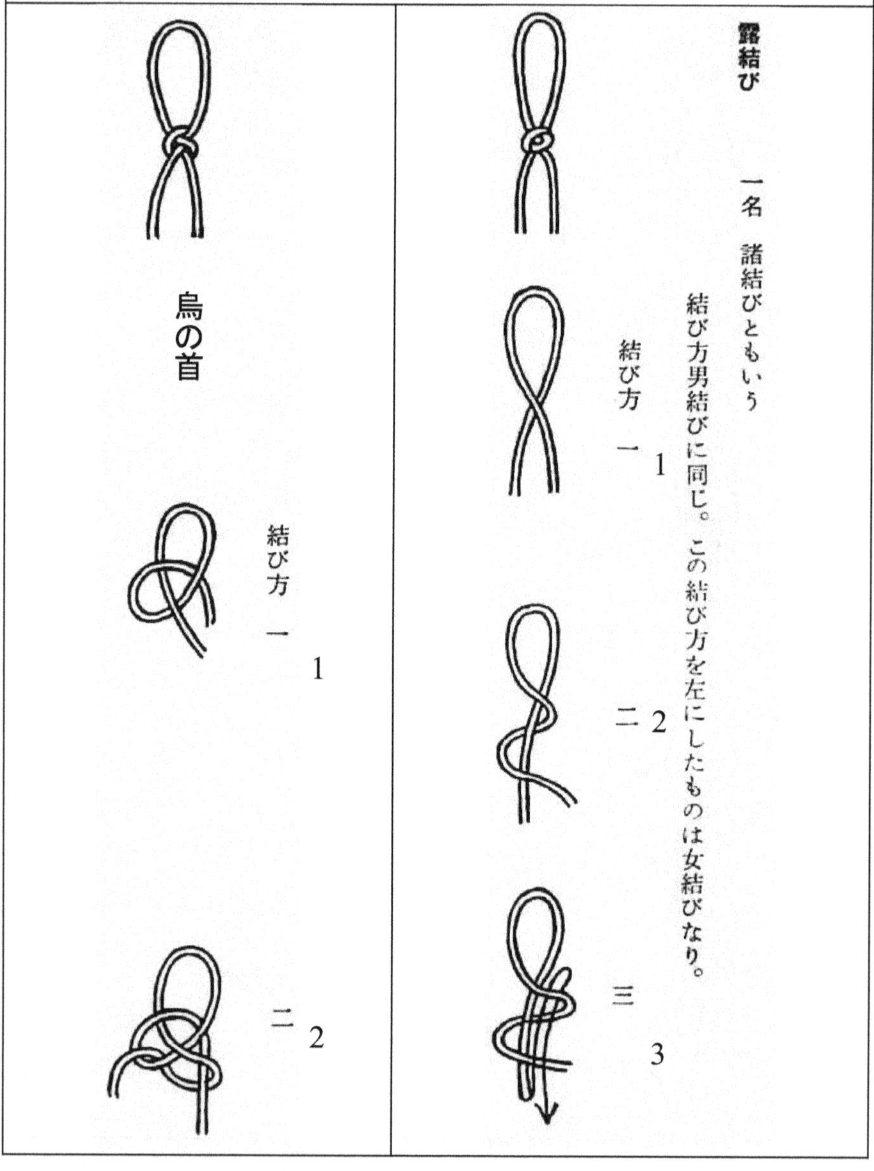

Musubi Kata 結び方 Knots	
How to tie Kanau Musubi: Grant Your Wish Knot	**How to tie Uto Musubi: Rabbit's Head Knot**
叶結び	兎頭結び
結び方 一 1	結び方 一 1
二 2	二 2

Musubi Kata 結び方 Knots	
How to tie U no Kubi Musubi: Cormorant's Neck Knot. This is also known as: Kake (?) Knot Kame Kugushi- Turtle Wrap Kamo Kuguchi- Duck Wrap Kamo Sage- Duck Restraint Wasagi Kashi Tsuke (?) Kama Gakushi- Hidden Sickle Intsuke- Mark Kame no Wa- Turtle Ring Knot	**How to tie Go Gyo Musubi: Five Element's Tie.** This tie is also known as Weaving Loom Knot. Note: The five elements are Earth, Fire, Water, Wood and Void (emptiness)

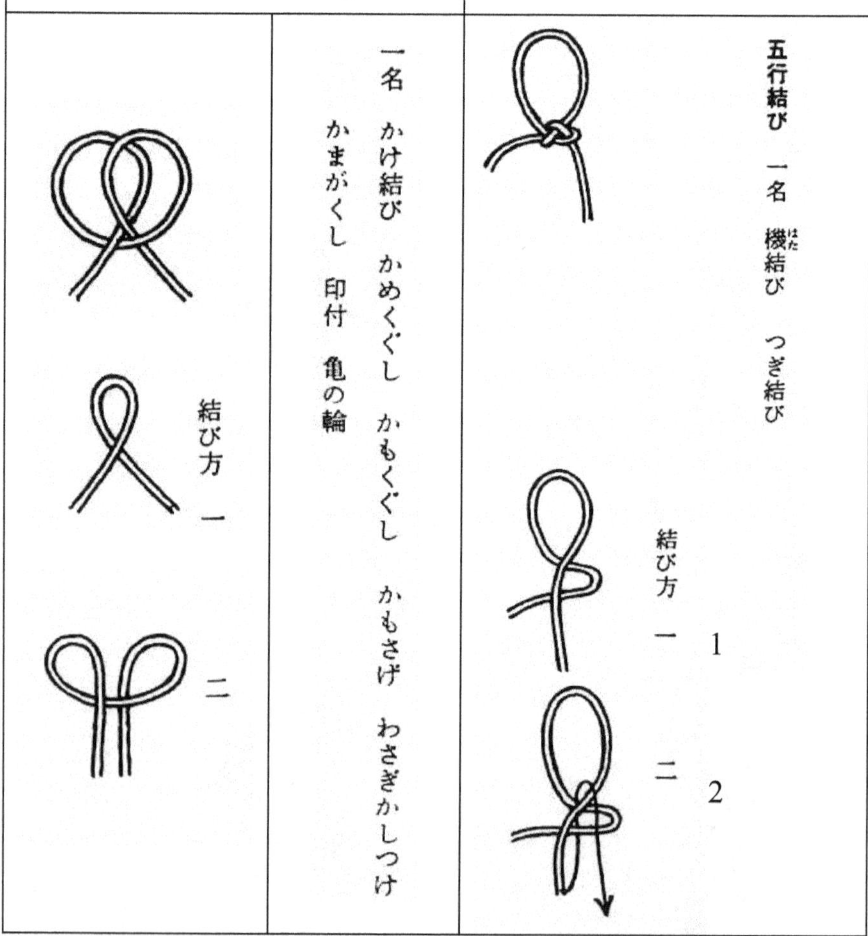

Musubi Kata 結び方 Knots

Kamo Sage: Duck Knot
Also called:
Kama Gakushi: Hidden Sickle
Kamo Washi: Duck Eagle
How to tie the Duck Knot is shown below.

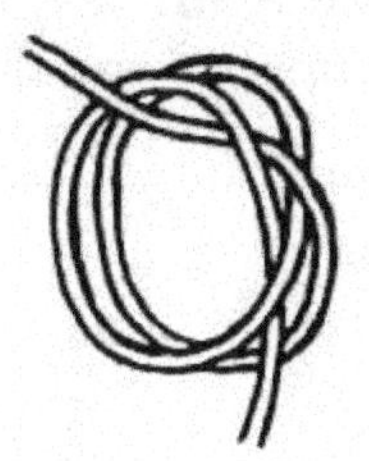

かもさけ
鎌がくし
鴨鷲

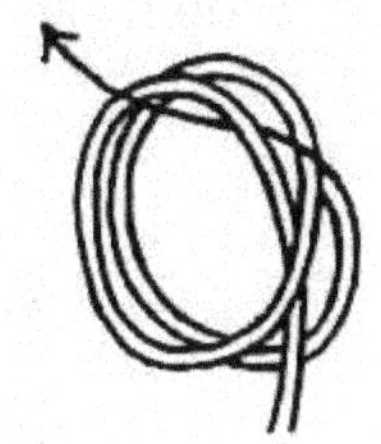

結び方

Inkai Hitoe Musubi
How to tie the Quick Pull Release One Layer Knot

Point A (イ) should pass through the loop in B（ロ）. Then pull tight. Then B should pass through the loop in A. The remaining rope should be tied off in a Men's Knot.

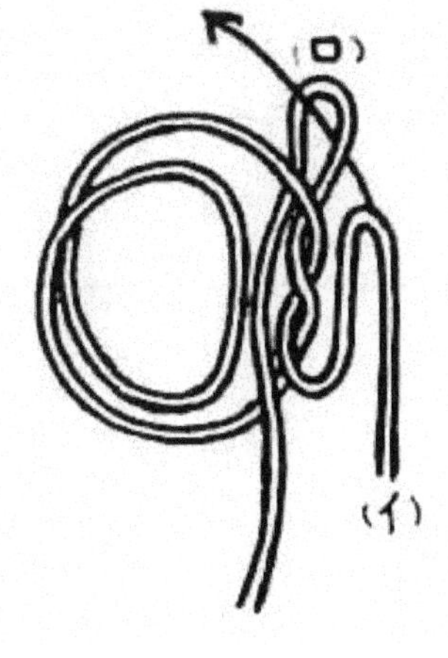

引解ひとえ結び

結び方
(イ)を(ロ)のわさに入れて引締め、さらにまた(イ)のわさに(ロ)をわさにして入れ残りの縄を男結びにして引き留める

Musubi Kata 結び方 Knots	
Tate Musubi: Vertical Knot This is also known as the Men's Knot	**Koma Musubi:** **Shogi Chess Piece Knot** This is also known as the Women's Knot
 **たて結び** 男結びともいう。 	 **こま結び** 女結びともいう。 

Musubi Kata 結び方 Knots		
Shin Musubi: True Knot	**How to tie Hata Musubi: Weaving Loom Knot**	
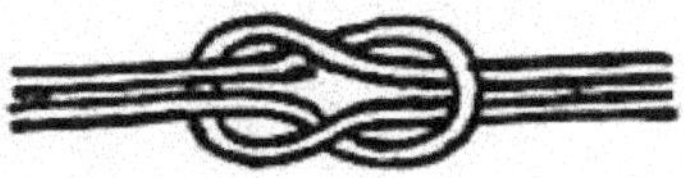	こま結び 玉結び こまか結び 結びきり 細結び 末結び 女結び 結びかけ	
This knot is also called: Koma Musubi: Shogi Chess Piece Knot Tama Mususbi: Ball Knot Komaka Musubi: Detail Tie Musubi Kiri: End of the Knot Hoso Musubi: Thin Knot Sue Musubi: End Knot Onna Musubi: Women's Knot Musubi Kake: Final Knot		

Musubi Kata 結び方 Knots	
How to tie U no Kubi Musubi: **Crow's Neck Knot**	**How to tie Hoju Musubi:** **Wish Fulfilling Jewel Knot** This knot is also known as an Omoi Musubi: Thinking Fellow's Knot or Rokuto Musubi: Six Headed Knot
 烏の首結び  結び方一 1  二 2 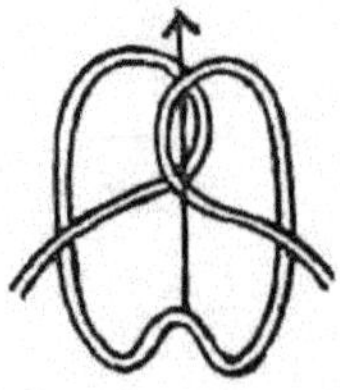 三 3	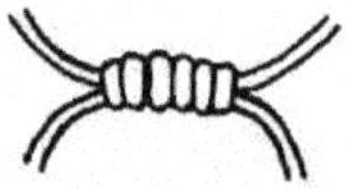 宝珠結び 思い結び 六頭結び  結び方一 1  二 2

Musubi Kata 結び方 Knots

How to tie Hishi Musubi: Water Chestnut Knot Also known as: Awabi Musubi: Abalone Knot Awaro Musubi: Awaro Region Knot Awazu Musubi: Unmatched Knot Tono Sama Musubi: Lord of the Domain Knot	**How to tie Nina Musubi: River Snail Knot** Kusari Musubi: Chain Knot
葵結び 蚫結び　淡路結び　不合結び　殿様結び 結び方 一 1 二 2	蜷結び 鎖結び おもて うら Front Back 結び方一 1 二 2 三 3

<table>
<tr><th colspan="3">Musubi Kata 結び方 Knots</th></tr>
<tr><td>結び方一 1
Pull out as arrows show
二 2
引く 引く</td><td>掛結び
一名 異掛帯結び 思結び 二つ華曼</td><td>How to tie Kake Musubi: Attaching Knot

This is also known as;
E-Kake Obi Musubi: Differing Belt Knot
Omoi Musubi: Thinking Fellow’s Knot
Futatsu Kaman: 2 Splendid Flowers</td></tr>
<tr><td> 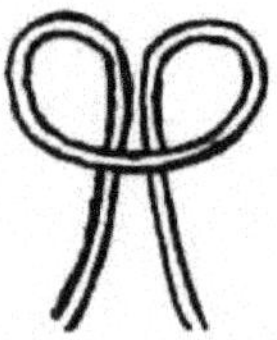雲雀結び</td><td colspan="2">Hibari Musubi: Skylark Knot</td></tr>
</table>

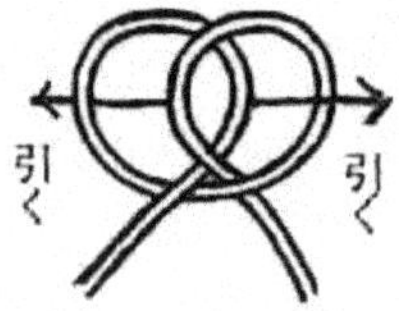

Musubi Kata 結び方 Knots

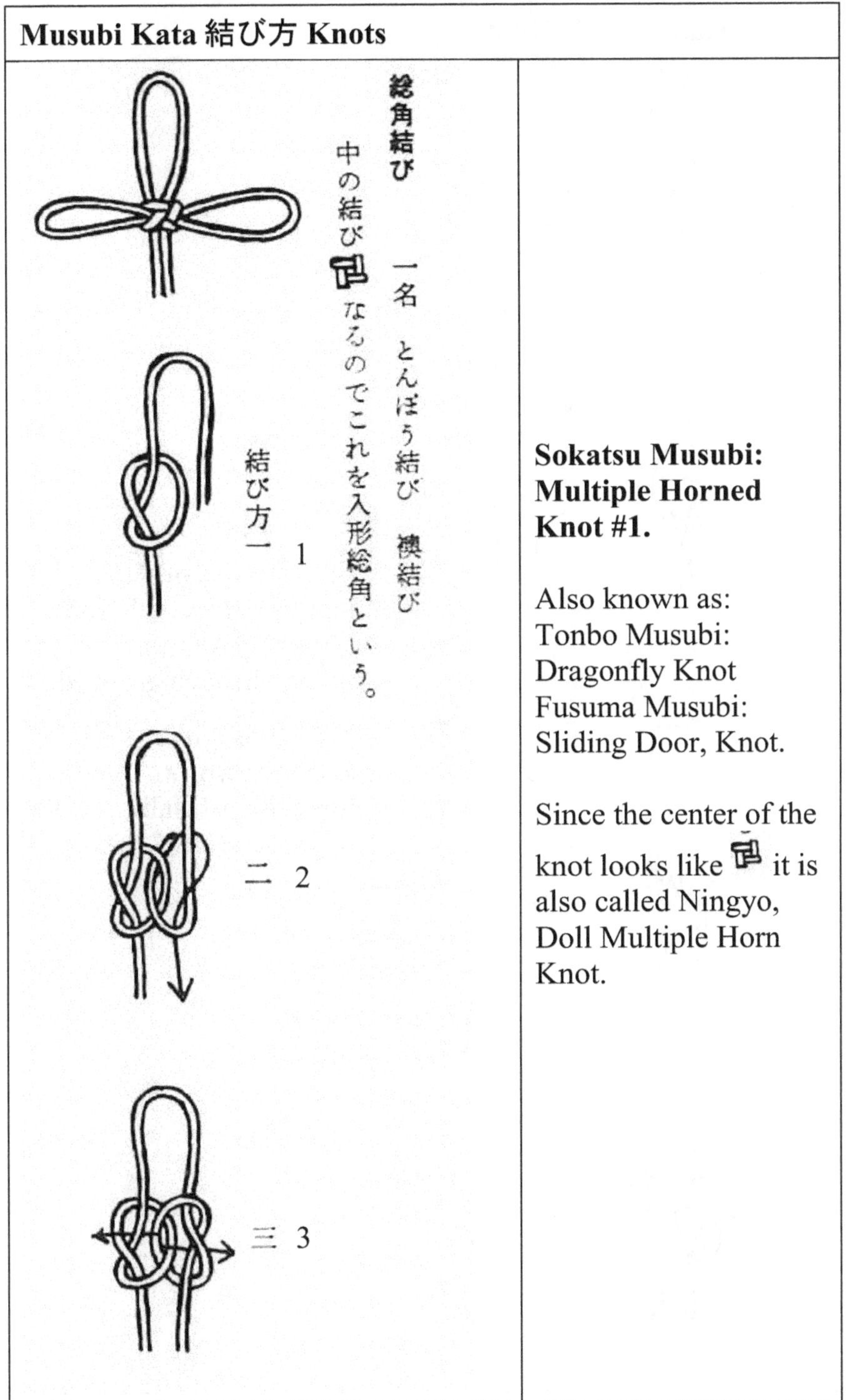

Sokatsu Musubi: Multiple Horned Knot #1.

Also known as:
Tonbo Musubi: Dragonfly Knot
Fusuma Musubi: Sliding Door, Knot.

Since the center of the knot looks like it is also called Ningyo, Doll Multiple Horn Knot.

Musubi Kata 結び方 Knots	
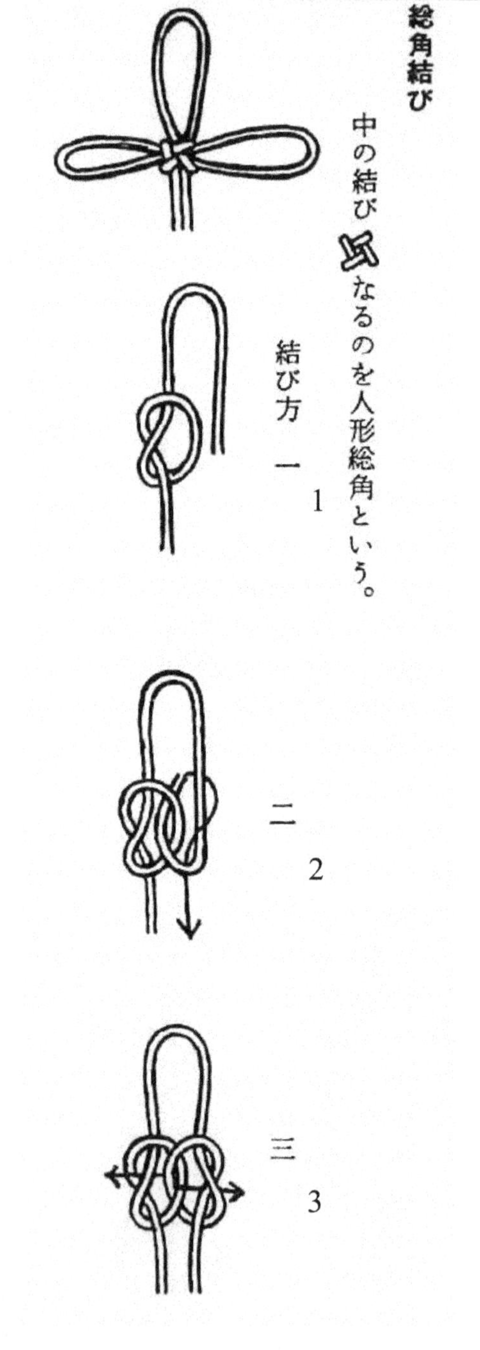 **総角結び** 中の結びなるのを人形総角という。 結び方 一 1 二 2 三 3	**Sokatsu Musubi: Multiple Horned Knot #2** Since the center of the knot looks like it is also known as Ningyo Sokaku Musubi, Doll Multiple Horn Knot.

Musubi Kata 結び方 Knots	
How to tie a Kake Obi Musubi **Inner Belt Knot**	**How to tie the Keman Musubi: Keman Knot** This knot is also known as Irihimo Dojin Musubi: Seeking Buddhist Enlightenment Knot.
掛帯結び 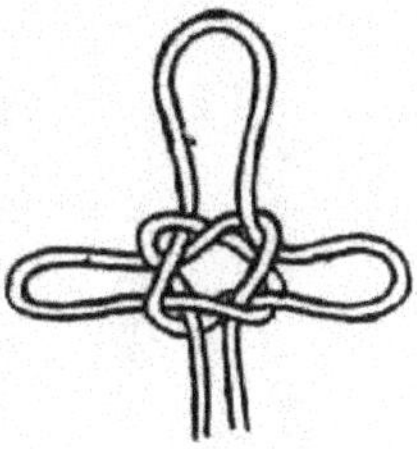 結び方 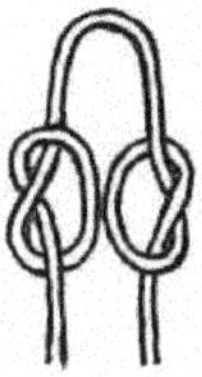 一 1 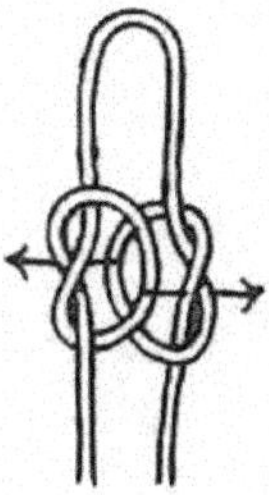 二 2	華曼結び 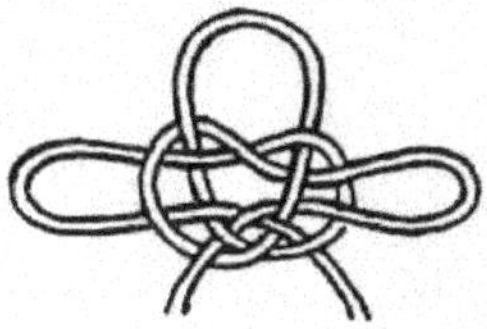 別名 入紐 道心結び 結び方  一 1 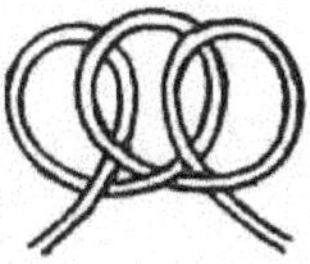 二 2  三 3

Musubi Kata 結び方 Knots
How to tie Tejo Nawa: Rope Handcuffs Also known as Rensa Musubi: Chain Tie Follow the instructions as shown in the illustrations

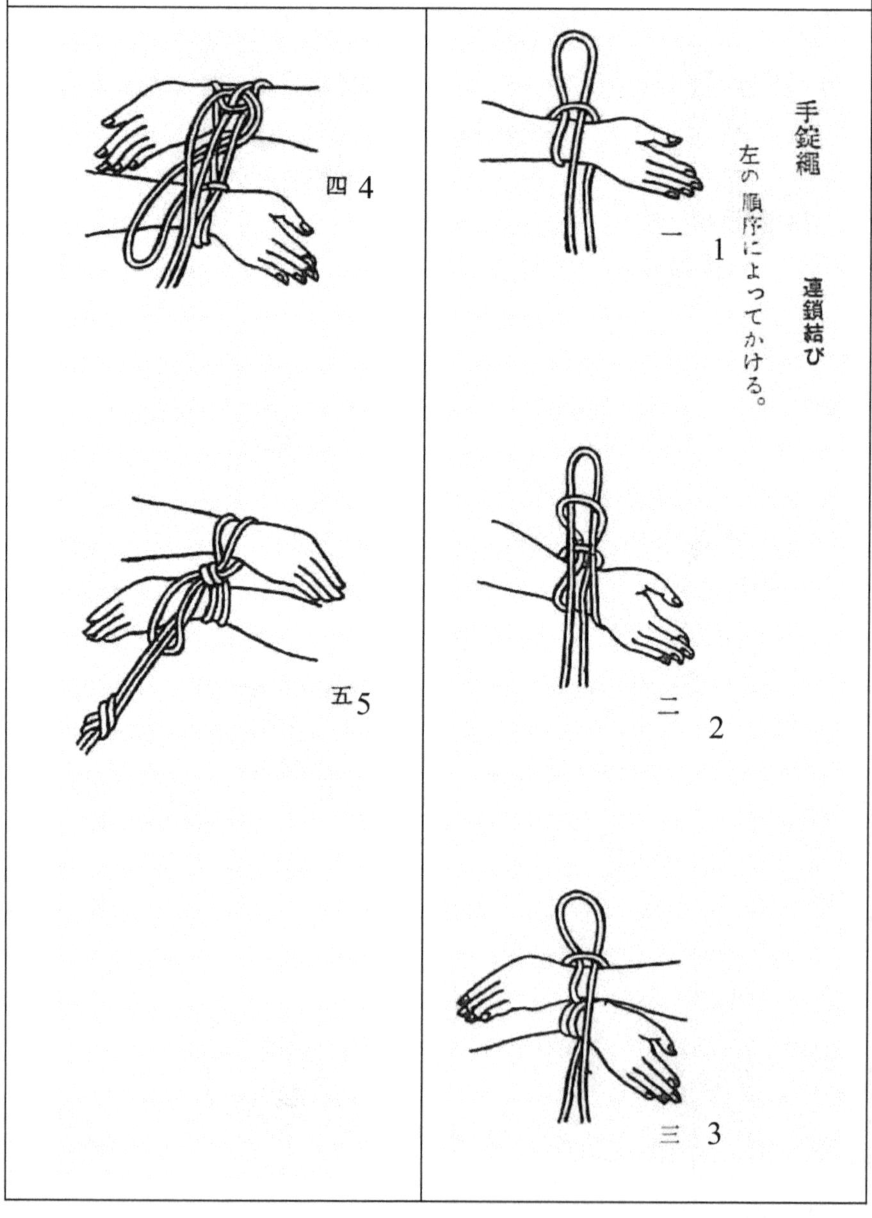

Musubi Kata 結び方 Knots

Tejo Nawa: Rope Handcuffs #1

This is also known as Dragonfly Tie. Follow steps 1, 2 & 3 to end up with the knot shown in 4. Place one of the prisoner's hands through each loop and pull tight.

1. Wrap the rope around the fingers three times.
2. Pull the ends of the rope through the middle
3. Put both prisoner's hands through and pull tight

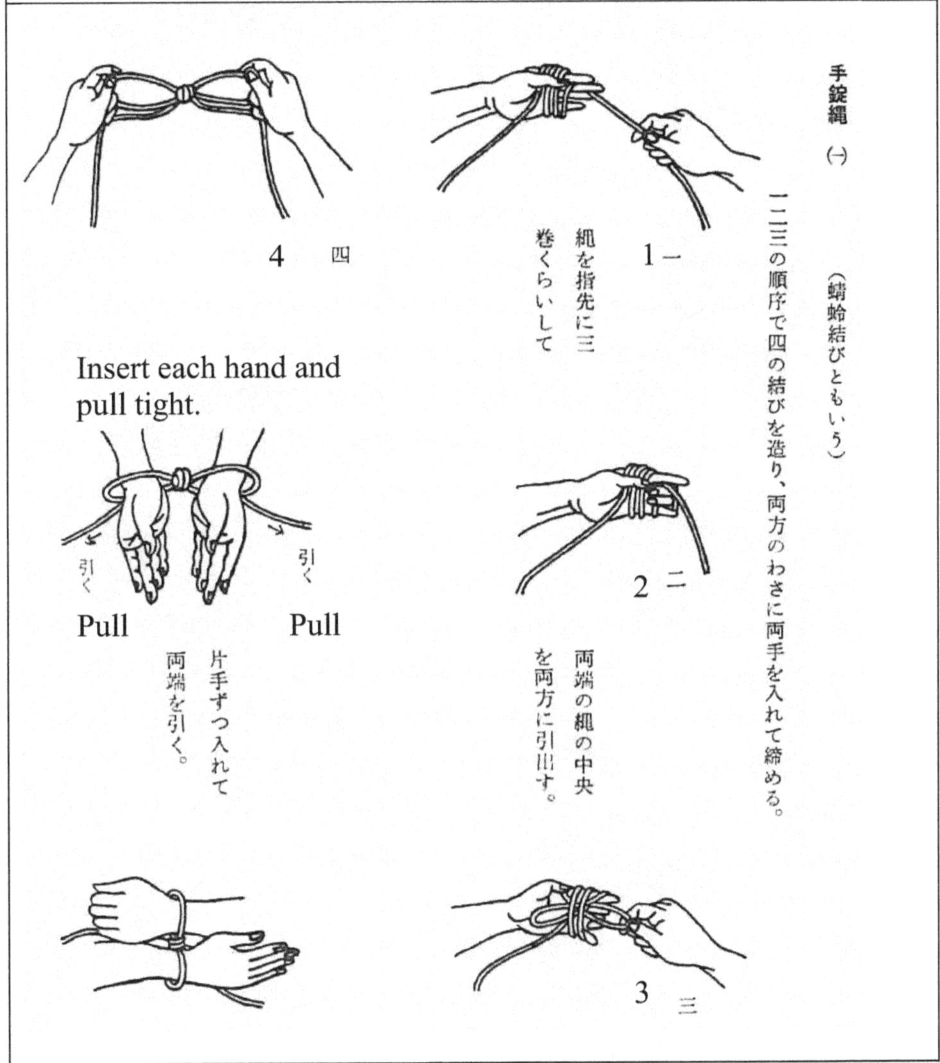

Musubi Kata 結び方 Knots	
Tejo Nawa #3: Rope Handcuffs #3	**Tejo Nawa #2:** Rope Handcuffs #2 This is a different way to tie #1

手錠縄

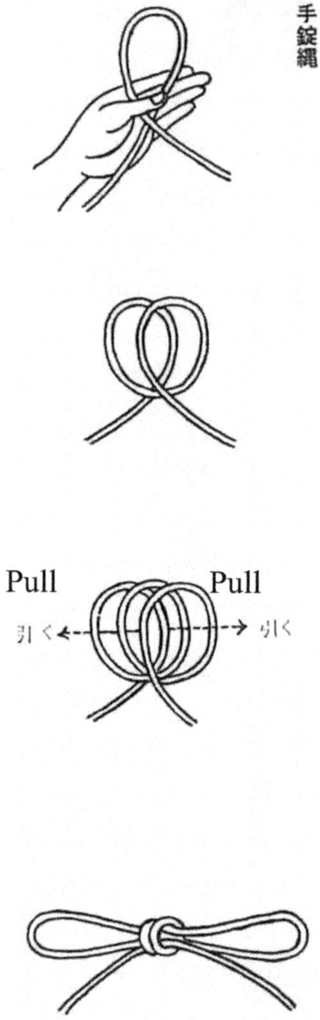

手錠縄

手錠縄(一)と同じ。

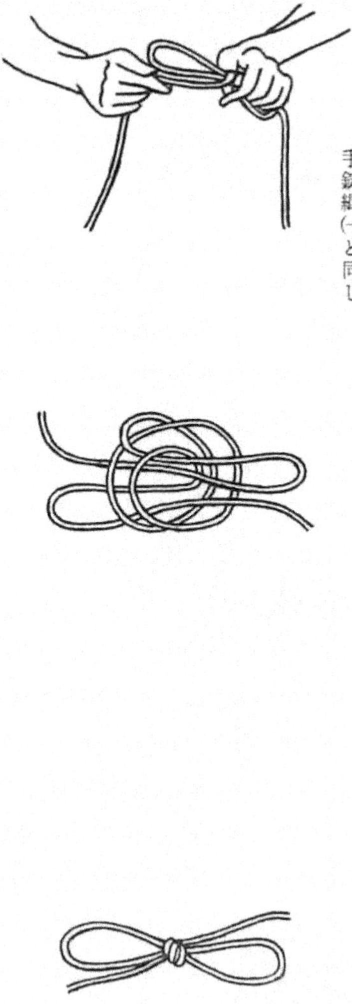

Pull 引く

Pull 引く

Musubi Kata 結び方 Knots	
Hiki Taki Musubi: Pull and Release Knot	**Hyo Musubi:** Gourd Knot
引き解き結び	瓢結び

Musubi Kata 結び方 Knots

Otoko Musubi: Men's Knot

1.

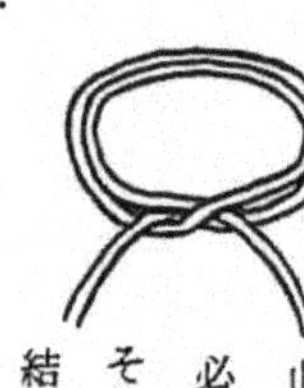

男結び

蜻蛉結び亀の輪等で縛ったその結び目は、その後の結び目の緩まぬように結び止めをする必要がある。それには男結びを用いる。

2.

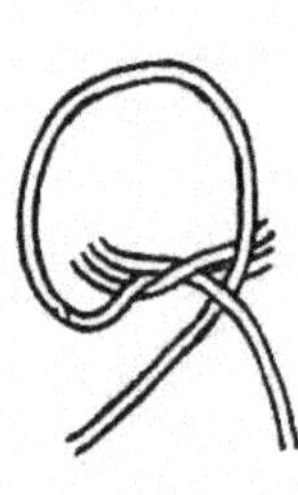

その結び方はどこでも締めた両端の一方を輪にして

3.

輪にしない一方を輪にした方に締めて差し込み、輪にした方の一方を引くと次のようになる。

4.

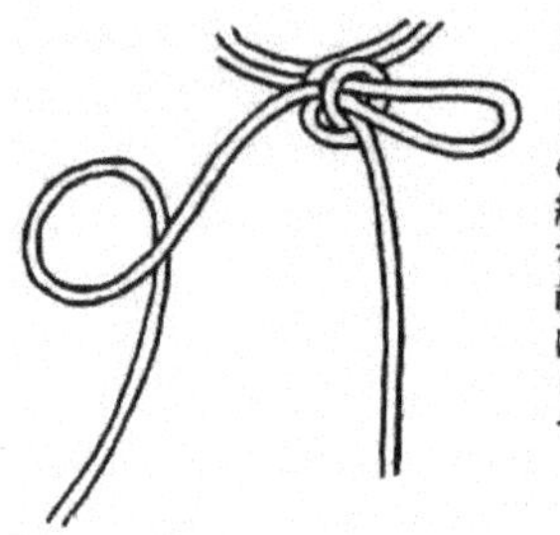

このようにしただけでは左の縄を引くとほどけるから、この縄を輪にして

5.

右方の結び目にかけおさえて左方に引くと

6.

このように結ばり、縄の両端いずれを引いてもほどけないようにする。

Musubi Kata 結び方 Knots	
Otoko Musubi: Men's Knot	
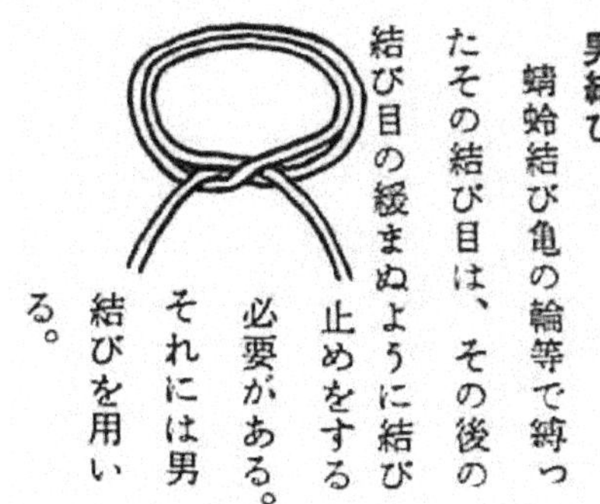	1. Use this after tying a Dragonfly Knot or a Turtle Ring Knot. This will ensure the knot tied above this one does not loosen. This is when the Men's Knot is used.
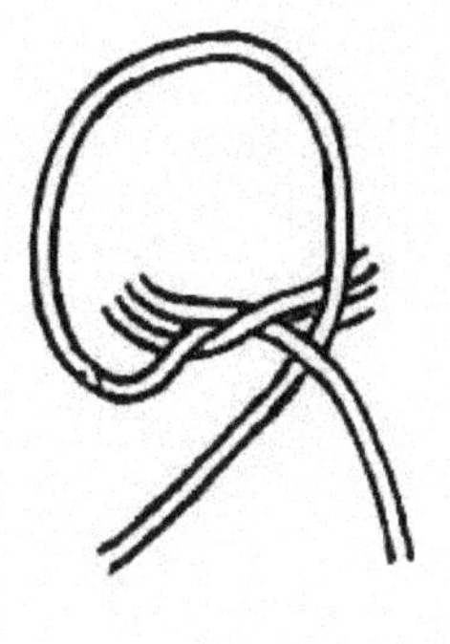	2. To tie this, take one of the ends of rope and form a loop.
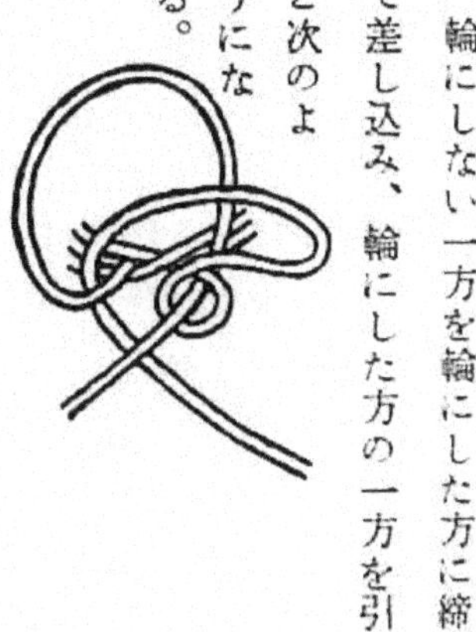	3. The end of rope you didn't form a loop with is passed through the loop. When you pull the first loop tight, it should look as shown in illustration #4.

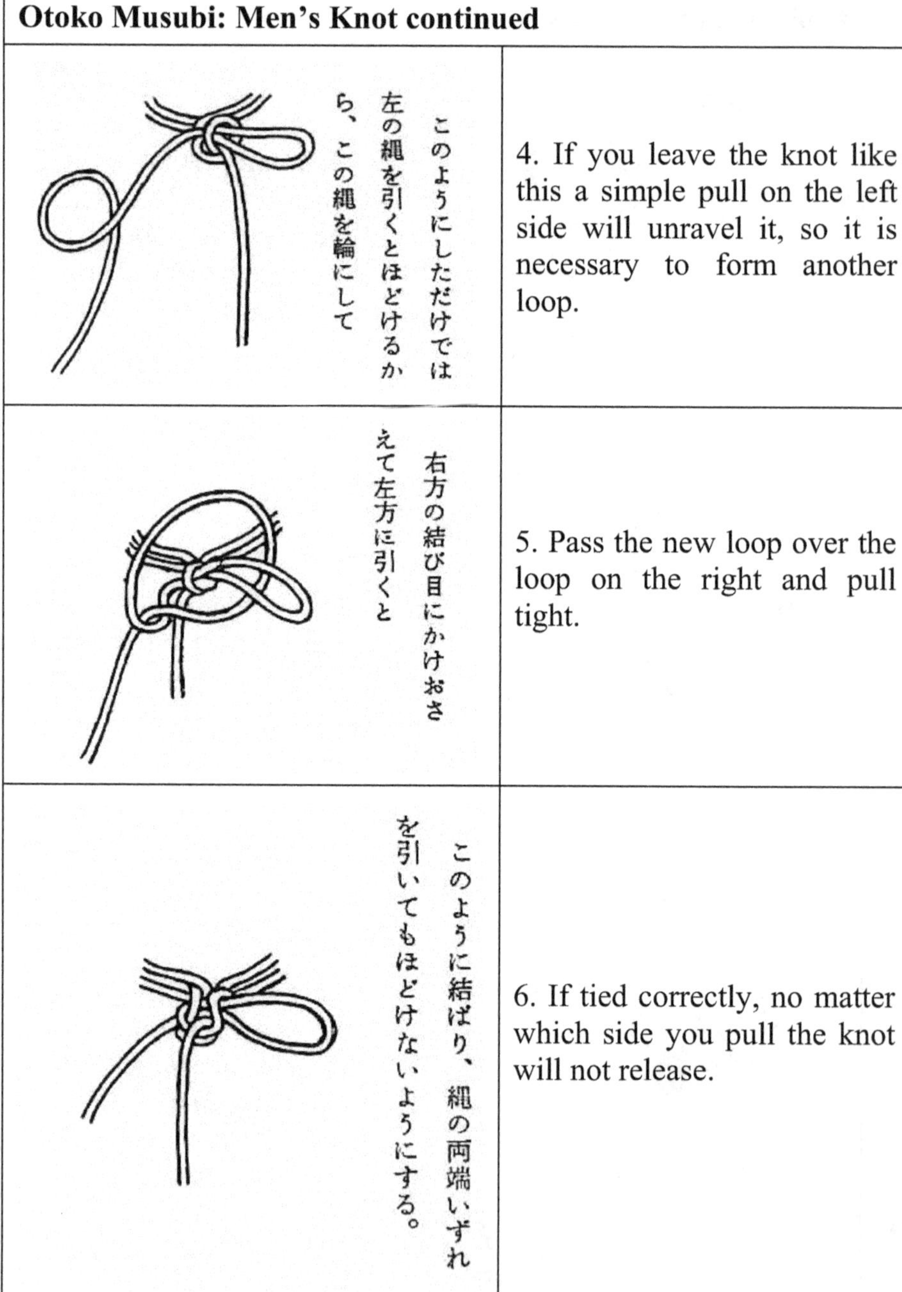

Musubi Kata 結び方 Knots	
Otoko Musubi: Men's Knot continued	
このようにしただけでは左の縄を引くとほどけるから、この縄を輪にして	4. If you leave the knot like this a simple pull on the left side will unravel it, so it is necessary to form another loop.
右方の結び目にかけおさえて左方に引くと	5. Pass the new loop over the loop on the right and pull tight.
このように結ばり、縄の両端いずれを引いてもほどけないようにする。	6. If tied correctly, no matter which side you pull the knot will not release.

Haya Nawa

早縄

Torite Yoh ・ Taiho Yoh

捕手用・逮捕用

Kakeyoh

掛様

How to Use Fast Tie for Capturing and Arresting

捕縄術には、早縄と本縄とがある。

早縄は速縄ともいい、また、仮縄、仮縛ともいう。乱棒狼藉を働く者を速時制禦拘禁したり、犯罪人をとりあえず逮捕する場合にかける捕縛縄で、一名捕手縄ともいう。

本縄は本式縄の略称で一名、本縛ともいった。すでに捕えた犯人を押送する場合、一定の縛縄方式によってかける縄縛法で、護送縄とも、っている。

この早縄、本縄を、流派によっては早縄本縄といわず、早縄は早縄でも、本縄を堅縄（かため）といっている流派もある。また、真行草に別け、真は本縄、草は早縄、行は本縄早縄いずれをも兼ねた縄縛法としているものもある。また、縄法に陰縄、陽縄というのがある。これはその縄掛けの方法を縄表より見て、上になるよう上からかけるのが陽縄、下になるようかけるのが陰縄である。

なお、縄形名称は流派によってそれぞれ違った名称が附せられ一定していない。同じような縄縛方法でも名称は多分に違っている。

Haya Nawa： Fast Tie

This knot is used to restrain captured enemies or criminals. There are two broad categories of Hojo binding:
Haya Nawa and Hon Nawa.

Haya Nawa, Fast Tie, is also known as Rapid Tie. Other names include:
Kari Nawa: Initial Tie
Kari Haku: Initial Restraint

This technique is used to rapidly subdue and maintain control over violent and deranged people. It also can serve as the initial restraints placed on a criminal under arrest. Other terms for it include Torite Nawa, Seizing Tie.

Hon Nawa: Main Tie

Hon Nawa, Main Tie, is an abbreviation of the full Name Honshiki Nawa, Formal Main Tie. It is also known as Hon Shibari, Main Restraint. It is used when a criminal has already been captured but needs to be transported to another location. This is a method of trying and restraining prisoners for transport. It is also known as Go So Nawa, Guarding and Transport Tie.

Depending on the Ryuha, or martial arts school, the names Haya Nawa and Hon Nawa may or may not be used. Some Ryuha use the name Haya Nawa but use the name Katame, or Lock Up, instead of Hon Nawa, Main Tie. Other schools divide them into 3 categories called Shin-Ko-So, True-Proceed-Grass. The 3 Kanji word Shin-Ko-So originated in China and described the three different ways to write the characters that make up the Chinese language. Later Shin-Ko-So was used in martial arts schools. Shin-Ko-So means Formally – Something-in-Between – Artfully. *Shin* represents the formal, prescribed method. *So* represents the artful/free application of the technique while *Ko*, is between the two.

Formally 真 *Shin* Represents Hon Nawa

Something in Betweeen 草 *So* Represents Haya Nawa

Artfully 行 *Ko* Hon Nawa and Haya Nawa used together.

The Yin and Yang duality also features prominently in Haya Nawa. Yin is darkness, the moon and feminine, while Yang is light, the sun and masculine. From the point of view of the observer it means if the person is upright then they are tied upright, thus representing Ying. If the person is on the ground when they are being bound, then it is representing Yang.

In the end the names of techniques vary according to the school and there are no rules. Even the same techniques can have different names in different schools.

Torikata: How to Arrest

Grab the end of the enemy's right hand and use your left arm to suppress the shoulder joint. This should keep the enemy flat on the ground. Quickly transfer the loop attached to the end of your arresting rope from your wrist onto the enemy's wrist.

Torikata: How to Arrest

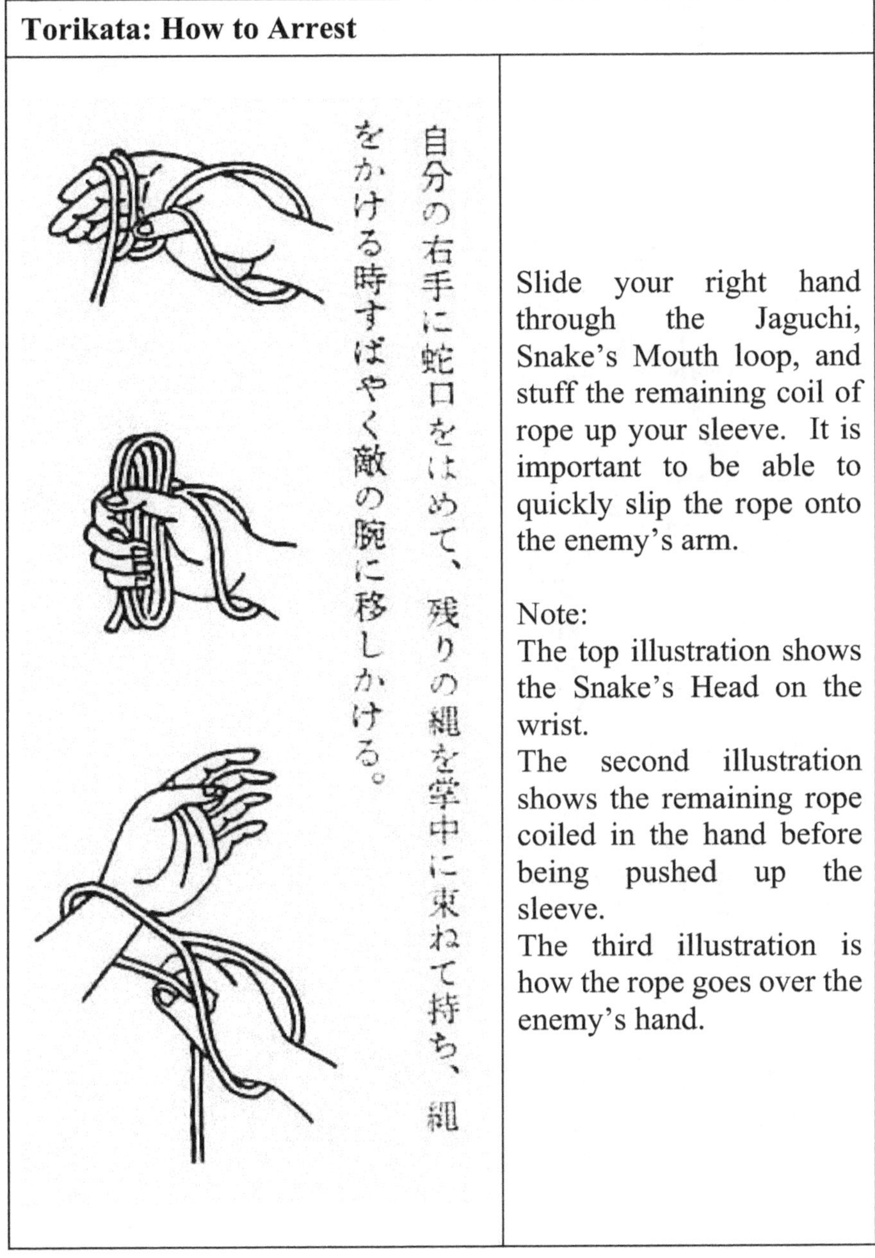

Slide your right hand through the Jaguchi, Snake's Mouth loop, and stuff the remaining coil of rope up your sleeve. It is important to be able to quickly slip the rope onto the enemy's arm.

Note:
The top illustration shows the Snake's Head on the wrist.
The second illustration shows the remaining rope coiled in the hand before being pushed up the sleeve.
The third illustration is how the rope goes over the enemy's hand.

Torikata: How to Arrest

With the Snake's Mouth around your wrist and the remaining rope coiled inside your sleeve, take hold of the enemy's wrist and transfer the rope over

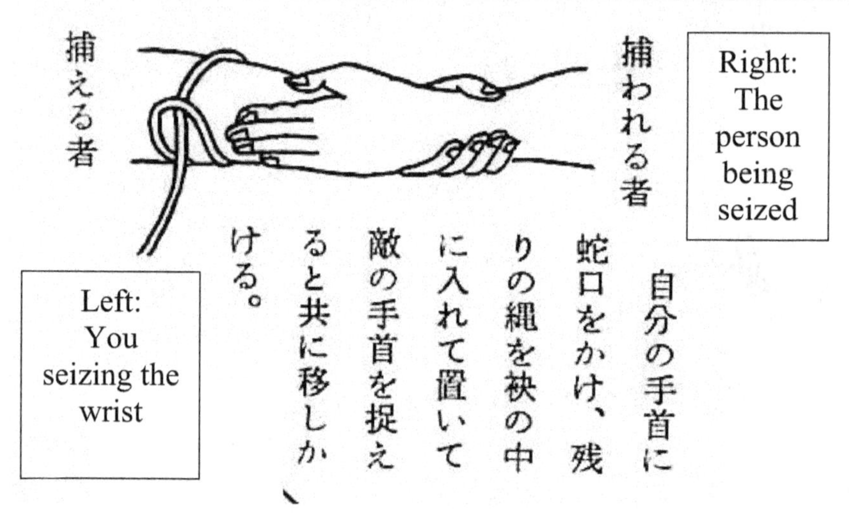

Note:
I believe the illustration got reversed since usually the rope is shown around the right wrist. I have corrected the illustration below.

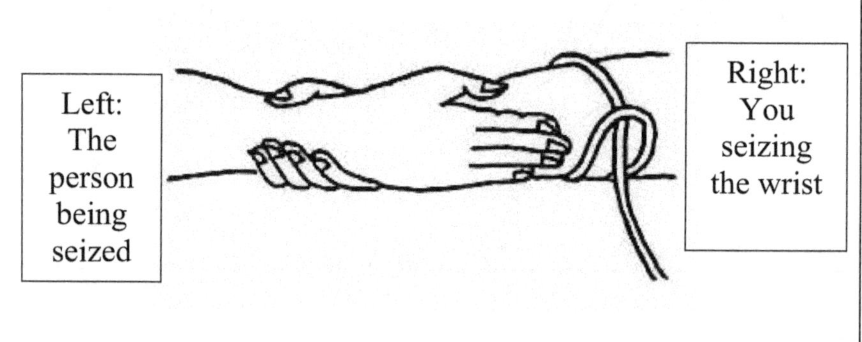

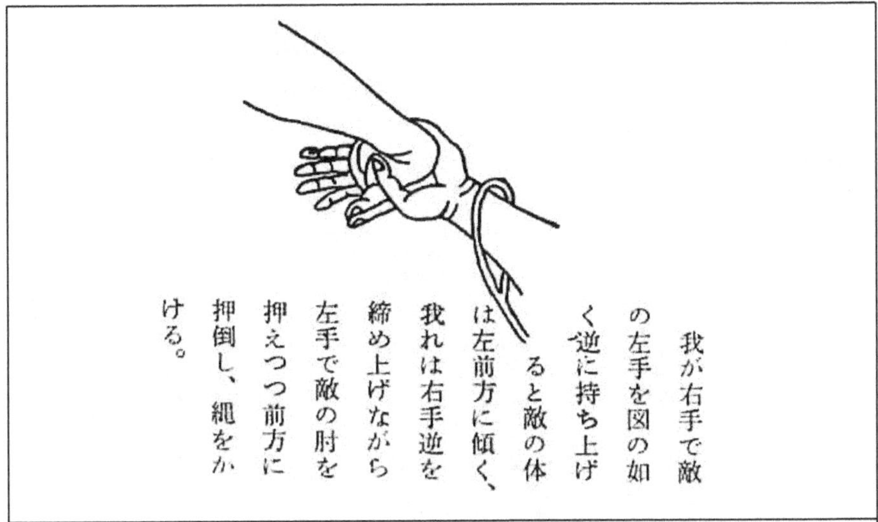

Take the enemy's left hand with your right hand as shown in the illustration. Apply a joint lock to his wrist and raise his arm. This will cause the enemy to lean forward and to the left. If you apply a wrist lock to the enemy's right wrist and raise it up while applying pressure to his right elbow with your left hand you will cause him to fall forward and you can then tie him with the arresting rope. This is shown in the illustration below.

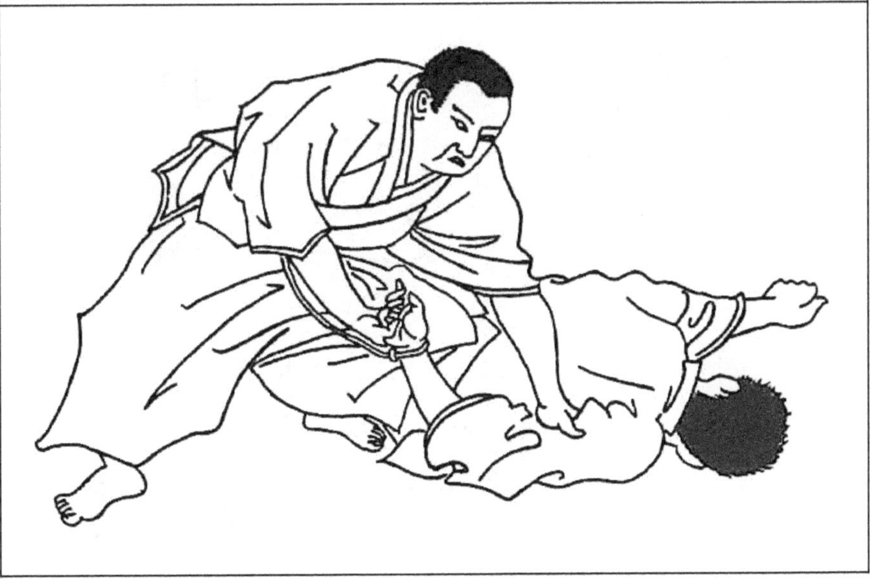

Shin Kage Ryu Haya Nawa: Fast Tie of the Shin Kage School
After attaching a loop to the enemy's right wrist, wrap the end of the rope around his throat from the left shoulder to the right. Bind his left arm and then wrap the rope twice around it and tie. Gather the remaining rope and tie it off 21 ~ 24 cm below the collar.

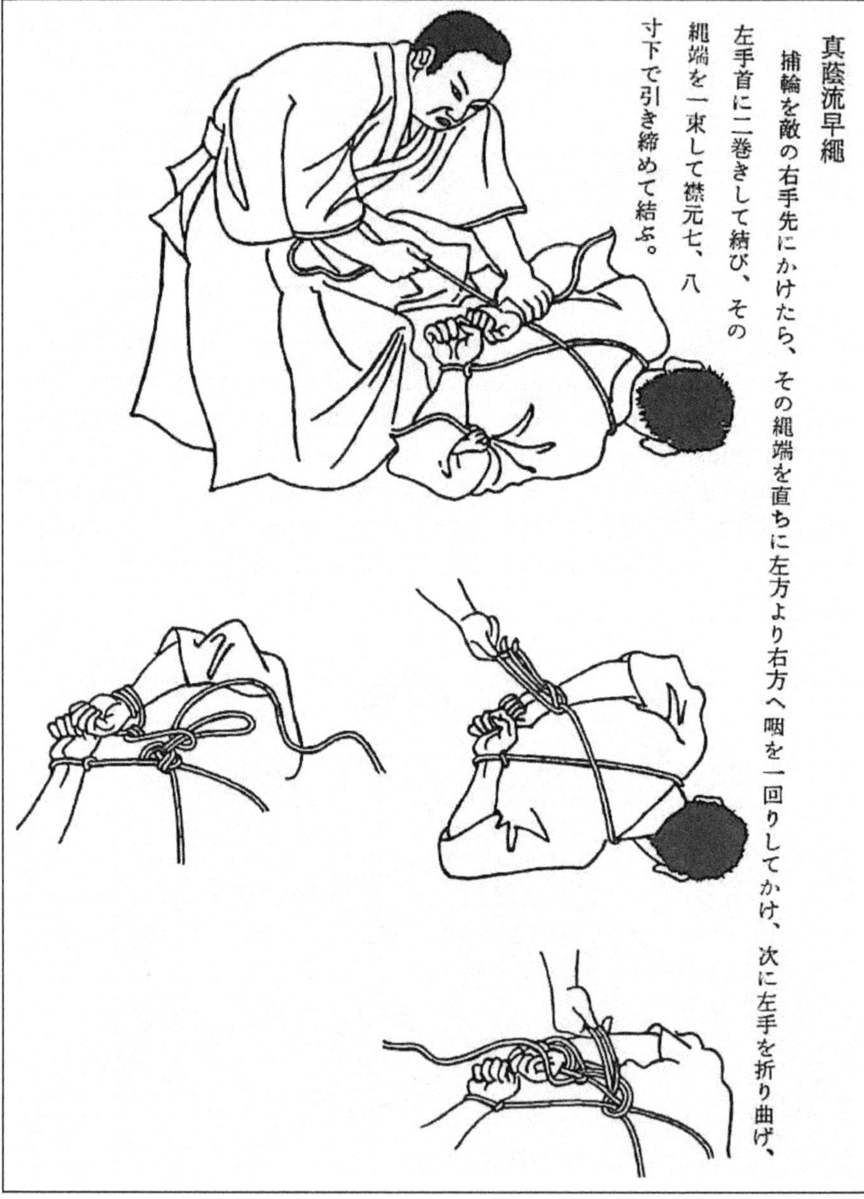

Torikata: How to Arrest

Sekiguchi Ryu Haya Nawa: Fast Tie of the Sekiguchi School

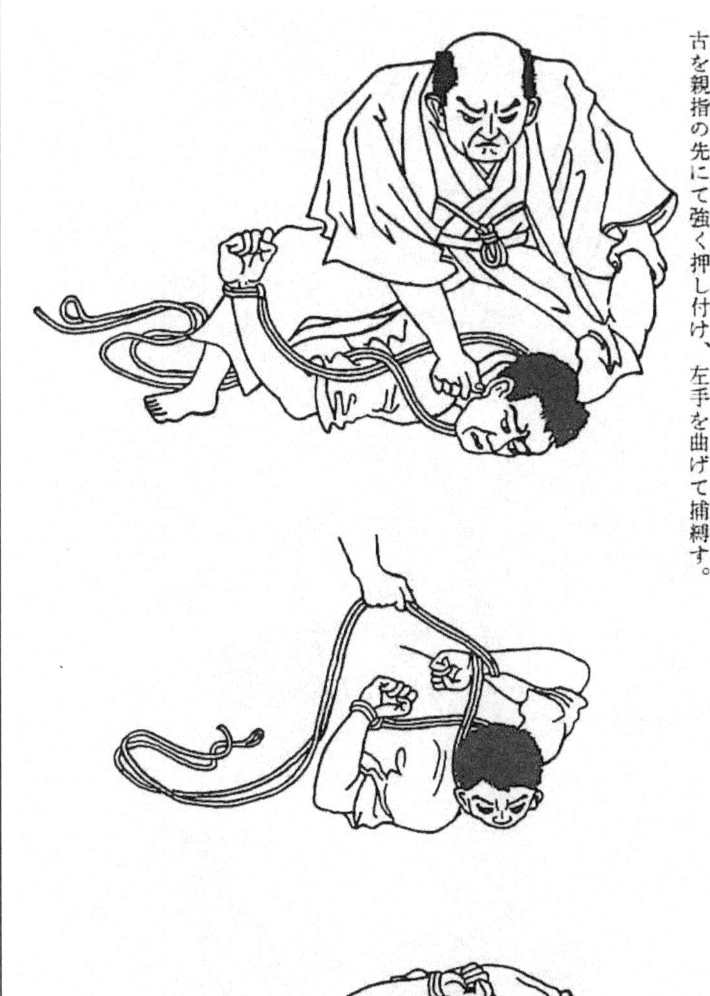

関口流早縄

捕縄を折半した真中を手首にかけ、縄端を左方より右方に首を一回りさして引き締め、左手を曲げて縄を手首に巻き付け、襟下八寸くらいのところで結ぶ。敵を倒し、その上に馬乗りに股がり、暴れる時は右耳下の急所独古を親指の先にて強く押し付け、左手を曲げて捕縛す。

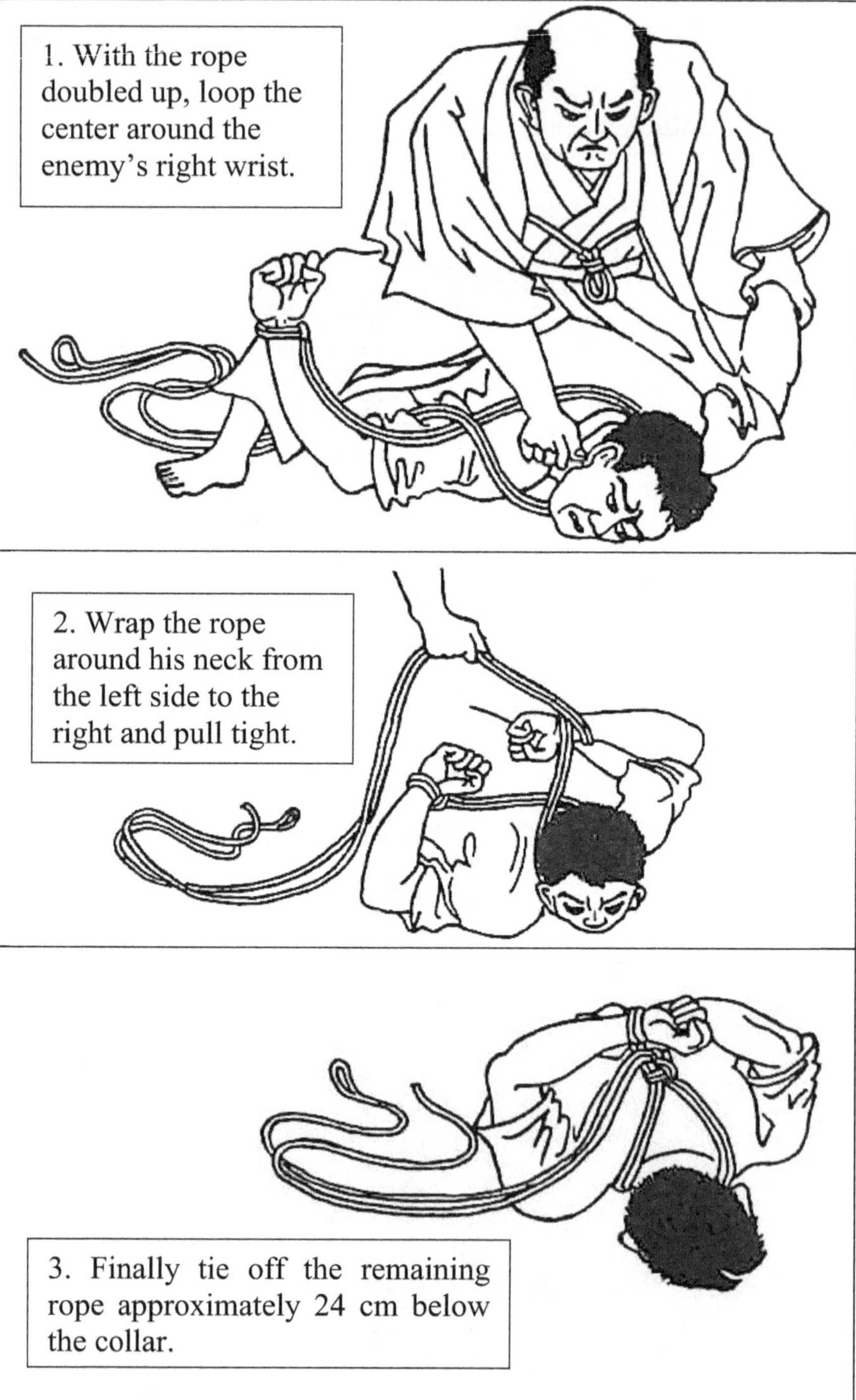
1. With the rope doubled up, loop the center around the enemy's right wrist.
2. Wrap the rope around his neck from the left side to the right and pull tight.
3. Finally tie off the remaining rope approximately 24 cm below the collar.

After you flatten the enemy on the ground straddle him. If he struggles then press hard with your thumb into the Kyusho, pressure point, known as Dokko, located just behind the earlobe. When you can bend the enemy's left arm, tie it off.

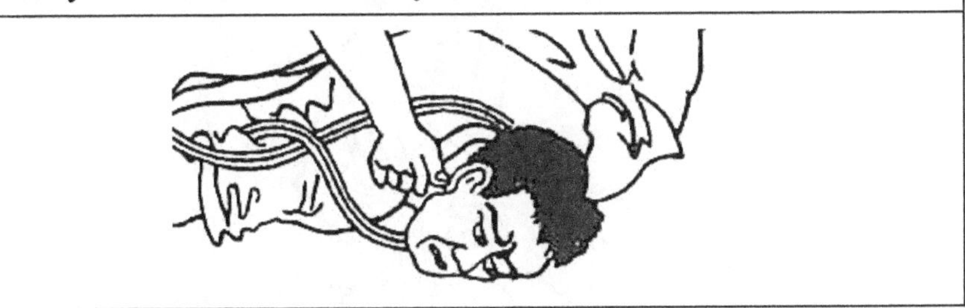

Note: The illustrations below show the point known as Dokko, which is used in Jujutsu based resuscitation methods. The illustrations below show how to use this point to reset a dislocated jaw. Left: *Jujutsu Striking Points and Resuscitation* by Fujimura Kinjiro 1895. Right: *Solo Jujutsu Practice Book* by Furugi Kennosuke 1911.

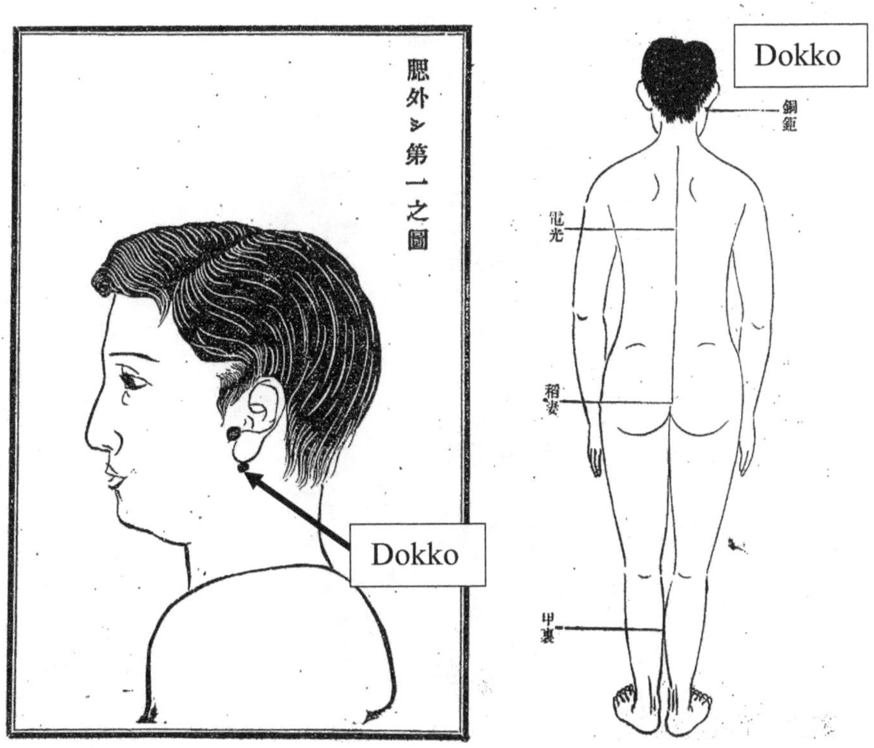

Tatsumi Ryu Haya Nawa: Fast Tie of the Tatsumi School

立身流早縄

敵を俯伏せに倒し、 左足にて二の腕を強く踏みつけ、 左手首にかけた縄端を左肩口より咽へ回して引き締め、左手を曲げてその手首に二巻回して結ぶ。 敵穏かなる時はそのままにしておいてよいが、 もし乱暴をなす時は直ちに縄の残りを以て左右の内の足の親指一本を結びつけて置く。

Torikata: How to Arrest

1. Bring the enemy down flat on his face. Step hard on his right Ni no Ude, bicep, and tie the rope around his right wrist.

2. Bring the rope over his left shoulder, around the throat and down over his right shoulder. Bend his left arm up and wrap it two times and tie off the remaining rope.

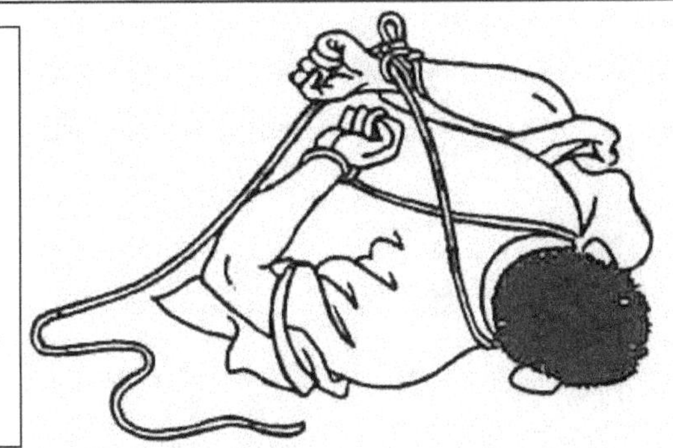

3. If the enemy is not putting up any resistance then the technique ends there. However, if he struggles then tie a loop around the big toe of either his left or right foot.

Note:
Fujita Seiko seems to have copied these illustrations from *An Illustrated Guide to Jujutsu Practice* 柔術練習図解 by Inoguchi Matsunosuke 1899. The illustration on this page is from that book.

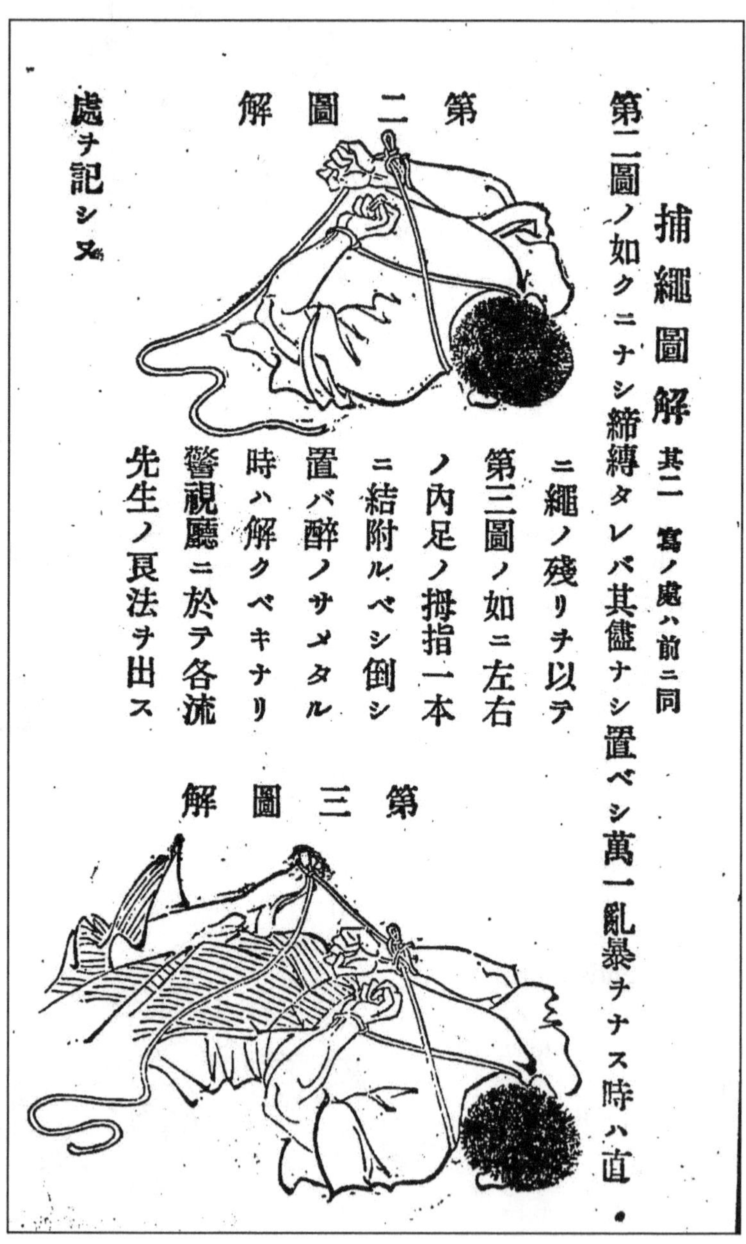

捕繩圖解 其二 寫ノ處ハ前ニ同

第二圖ノ如クニナシ締縛タレバ其儘ナシ置ベシ萬一亂暴ヲナス時ハ直ニ繩ノ殘リヲ以テ第三圖ノ如ニ左右ノ內足ノ拇指一本ニ結附ルベシ倒シ置バ醉ノサメタル時ハ解クベキナリ警視廳ニ於テ各流先生ノ良法ヲ出ス處ヲ記シヌ

第二圖解

第三圖解

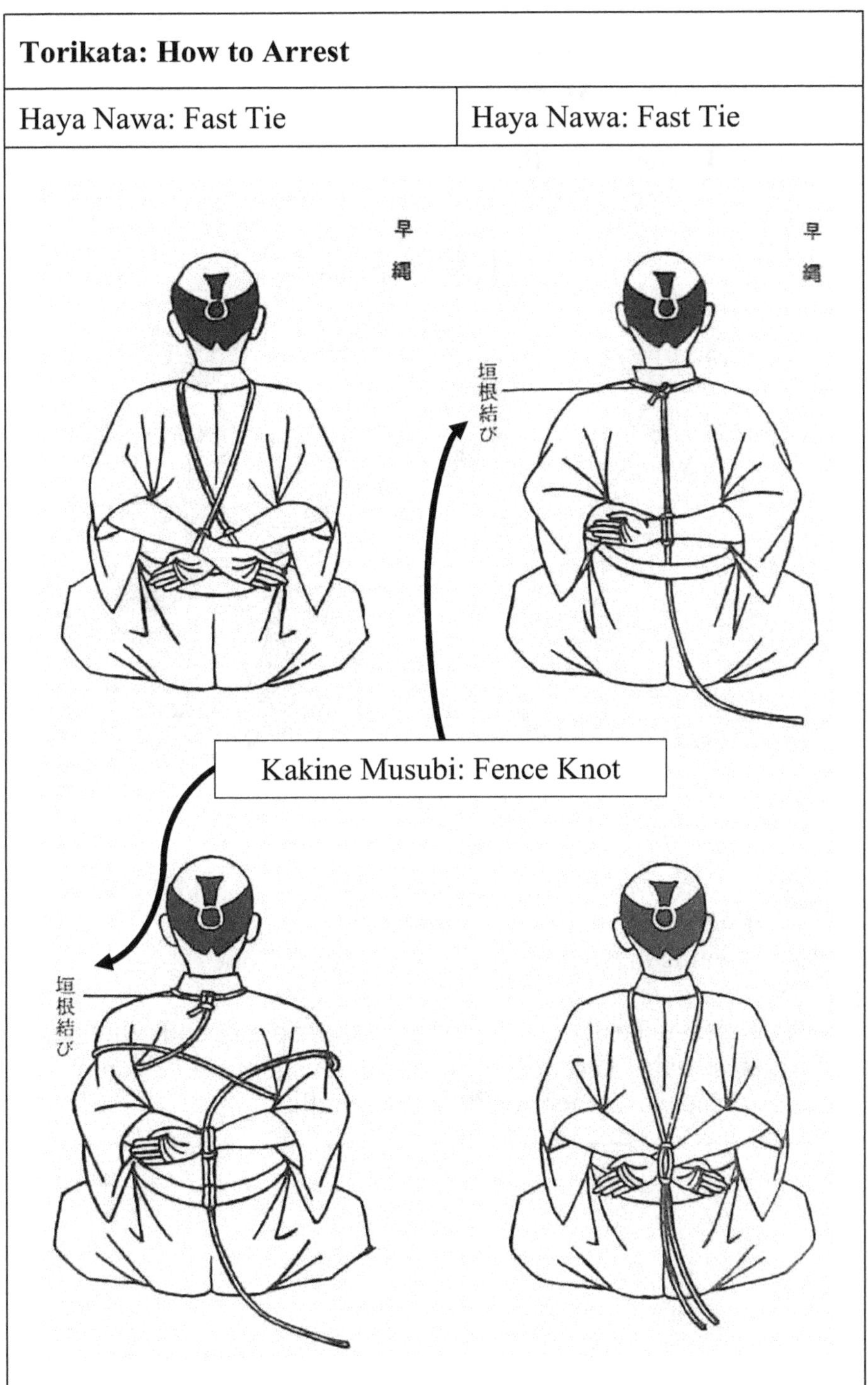
Torikata: How to Arrest
Haya Nawa: Fast Tie
Haya Nawa: Fast Tie
早縄
早縄
垣根結び
Kakine Musubi: Fence Knot
垣根結び

Torikata: How to Arrest
Motoyui Tome: Hair Binding String Tie #1

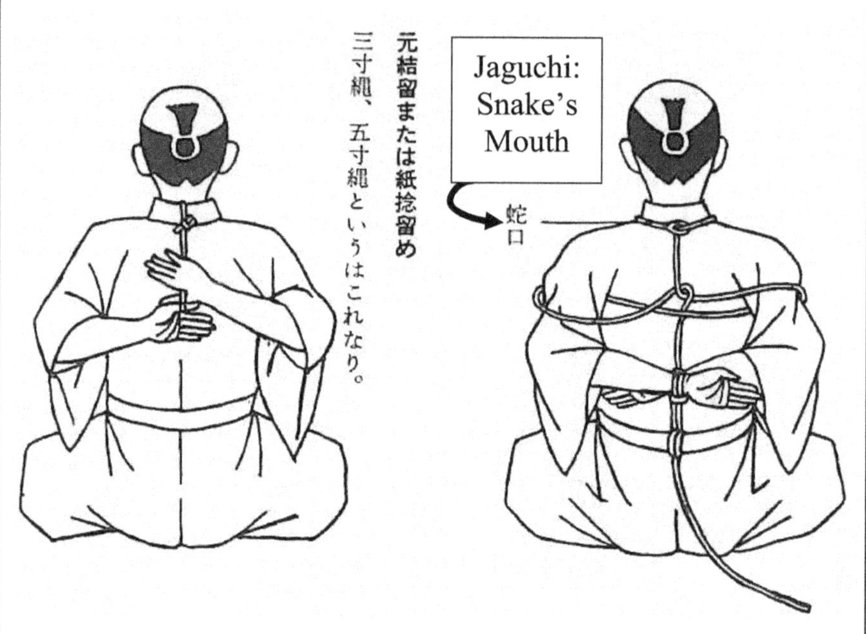

Motoyui Tome: Hair Binding String Tie
Kami Hineri: Twisted Paper String Tie
This is what is meant by "9cm" and "15cm" Rope

Note: Motoyui string is used to tie up the topknot and other parts of Japanese hair. The string was readily available due to common use. A piece was generally 70 cm in length.

Torikata: How to Arrest
Motoyui Tome: Hair Binding String Tie #2
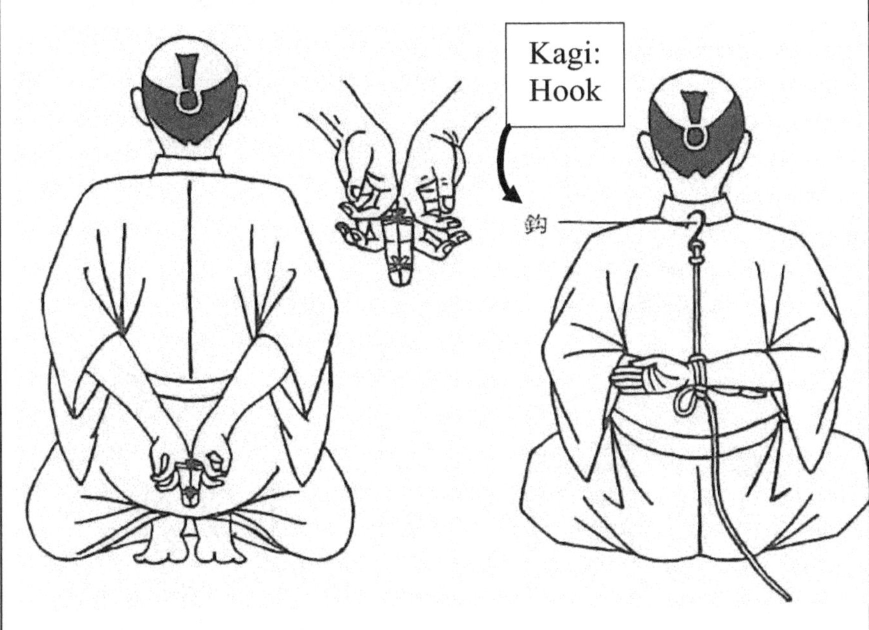
Detail of finger tie
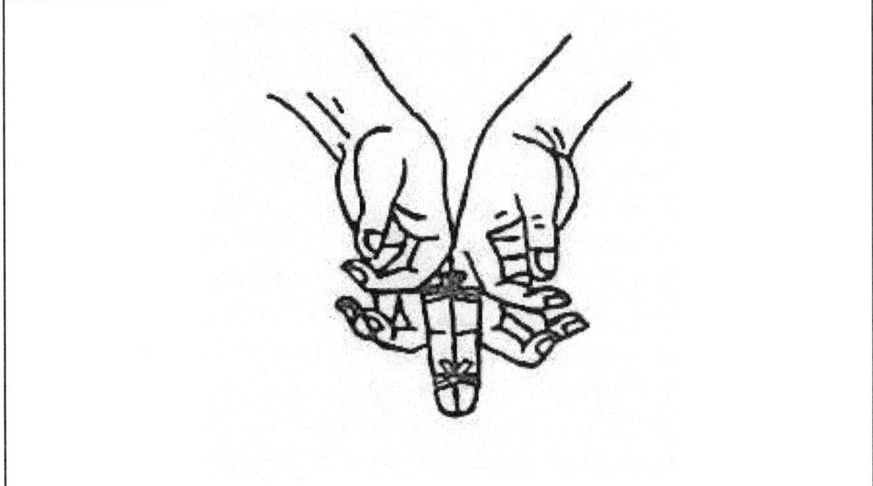

Torikata: How to Arrest	
十文字 Jumonji: **Cross Shaped**	**Koyori Tome:** **Paper String Finger Cuffs** Use a piece of Koyori, string made from twisted paper, to tie off both thumbs at the joint they meet the palms.
	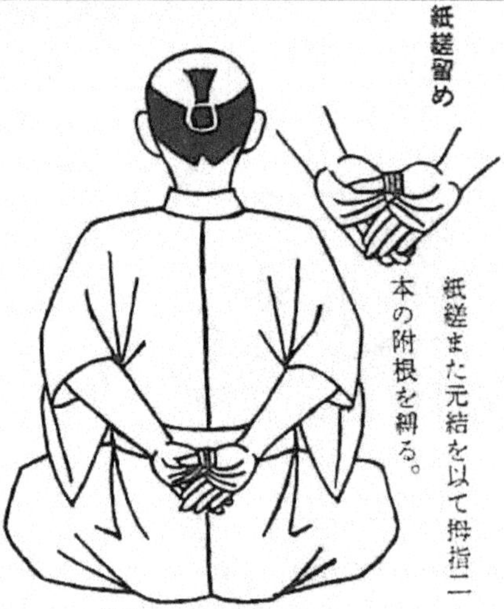 紙縒留め 紙縒また元結を以て拇指二本の附根を縛る。

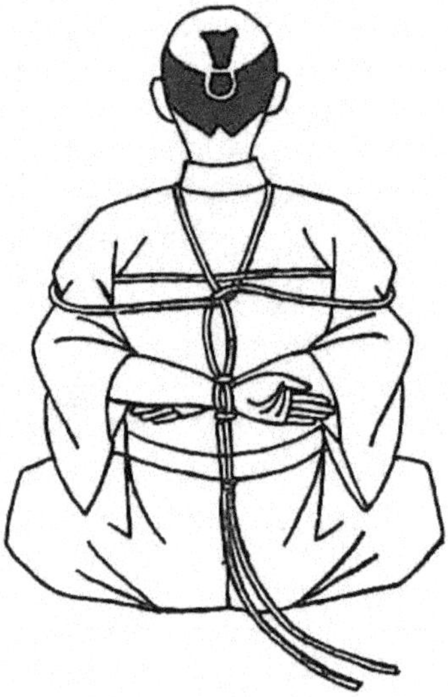

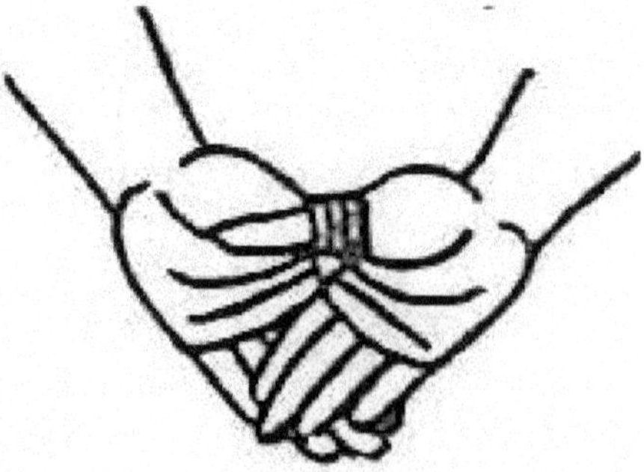

Torikata: How to Arrest	
Kote Kagi: **Wrist Hook**	**Hishi: Caltrop Tie** **Mawashi Nawa: Wrapping Around Tie** If you wrap the rope around the arms from above it is known as Yo Hishi or Yin Caltrop. If you attach it from the bottom it is In Hishi, Yang Caltrop. Note: In English the Chinese pronunciation of Yin and Yang is well known. In Japanese the pronunciation is In and Yo.
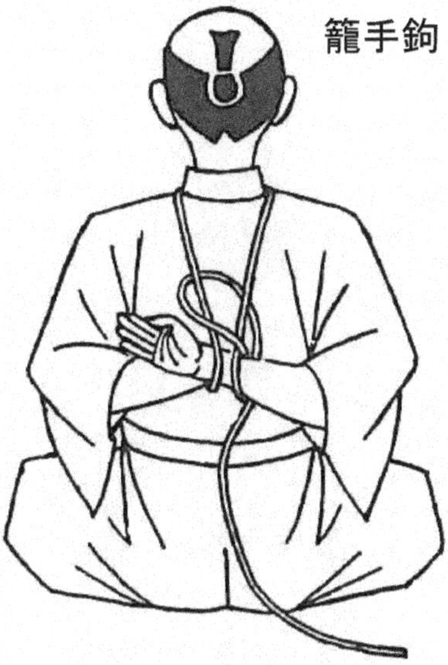 籠手鉤	 菱ともまわし縄ともいう 腕のところにかける縄の上よりかけるを陽菱といい、下にするを陰菱という。

Torikata: How to Arrest
Kagi Nawa: Hook Tie The hook is attached to the Eri, or front of the collar.

Torikata: How to Arrest	
Kuzushi Ryote Tori: Broken Two Handed Tie 崩し両手捕	Tasuki Tori: Sleeve Cord Tie 襷捕

Ushiro Kate-te Nawa: Back One-Handed Tie 後片手縄	Ushiro Tejo Nawa: Back Rope Handcuffs 後手錠縄
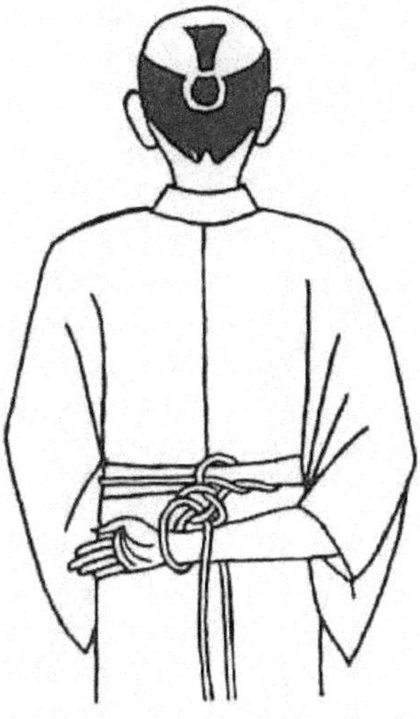	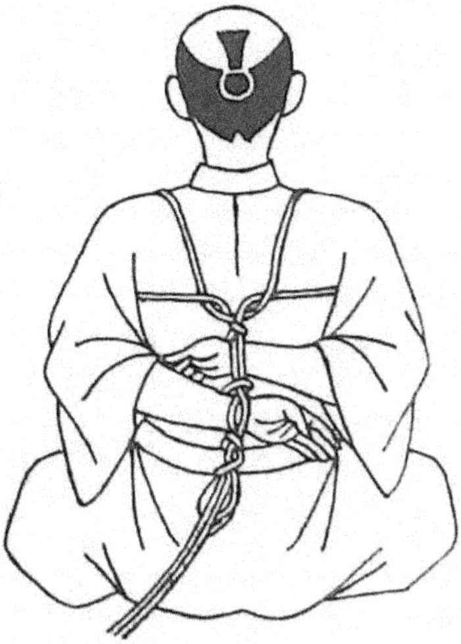

腰縄 **Koshi Nawa: Waist Tie ending with Men's Knot 1 ~ 7**

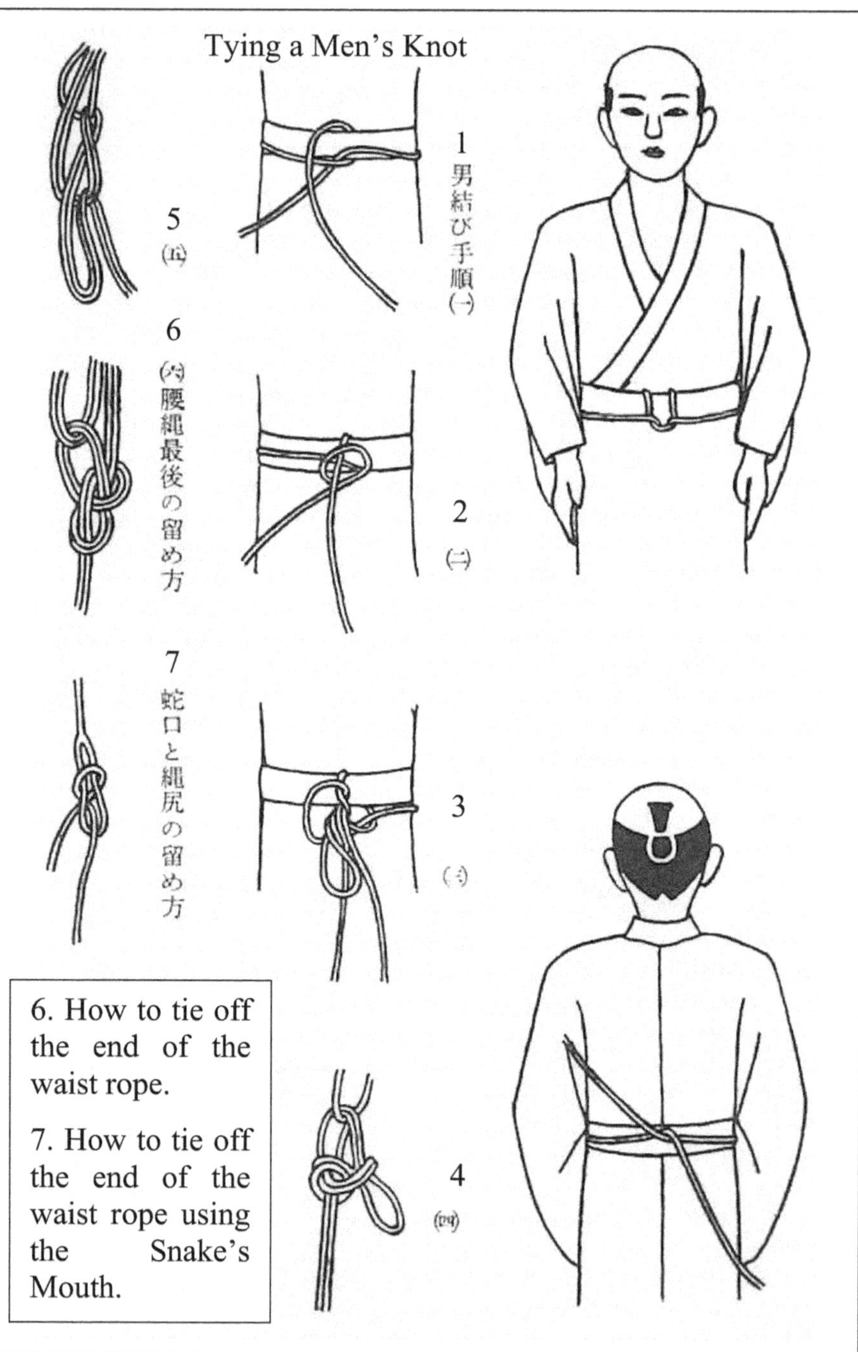

Torikata: How to Arrest

引致縄（連鎖縄）これは数人の者を同時に制縛する連鎖縄の方法で、これには蜻蛉結びを利用するのと、亀の輪を利用するのとの二通りある。

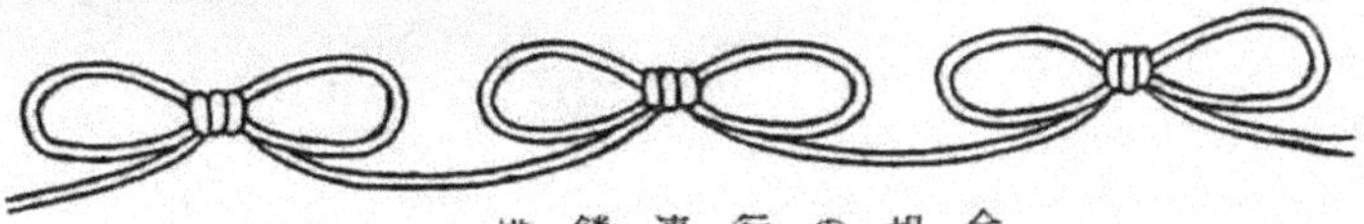

横鎖連行の場合

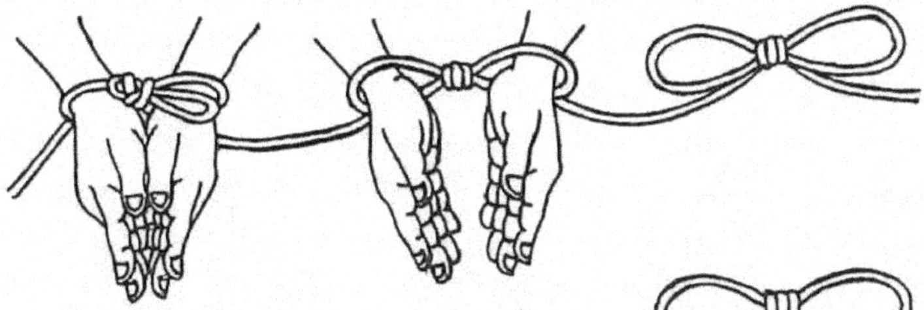

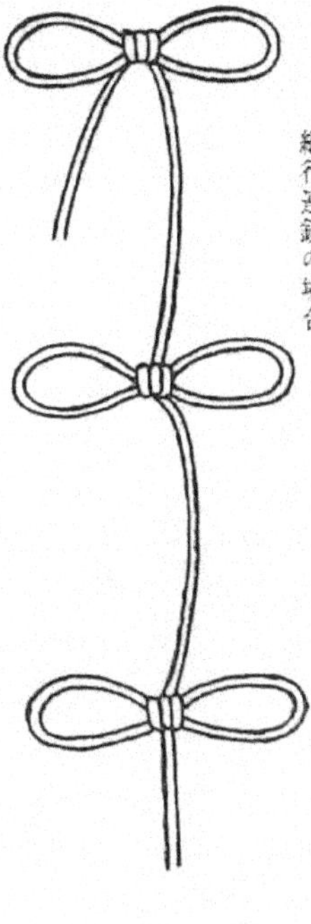

縦行連鎖の場合

蜻蛉結びを用いて一人ずつ縛する場合でも、亀の輪を用いて縛する場合でも、一連ごとに必ずその中央を一巻二巻してしっかりと締め、これを男結びで止めてから、次々と施縄するようにせねばならない。

蜻蛉結び

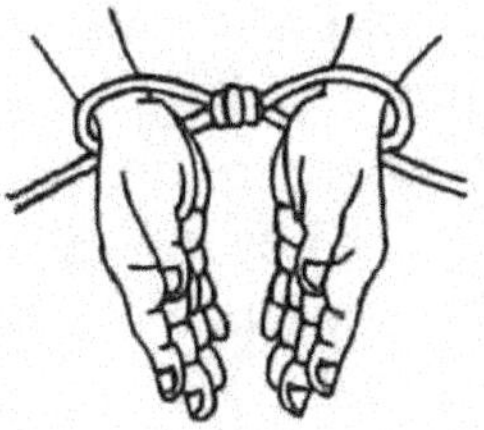

Torikata: How to Arrest

Inchi Nawa: Arrest Tie
This is also known as Rensa Nawa, Chain Rope. You can use this when arresting multiple people. You can use either Dragonfly Knot or Turtle Ring Knot.
1 & 2: How to tie in a horizontal line
3 : How to tie in a vertical line

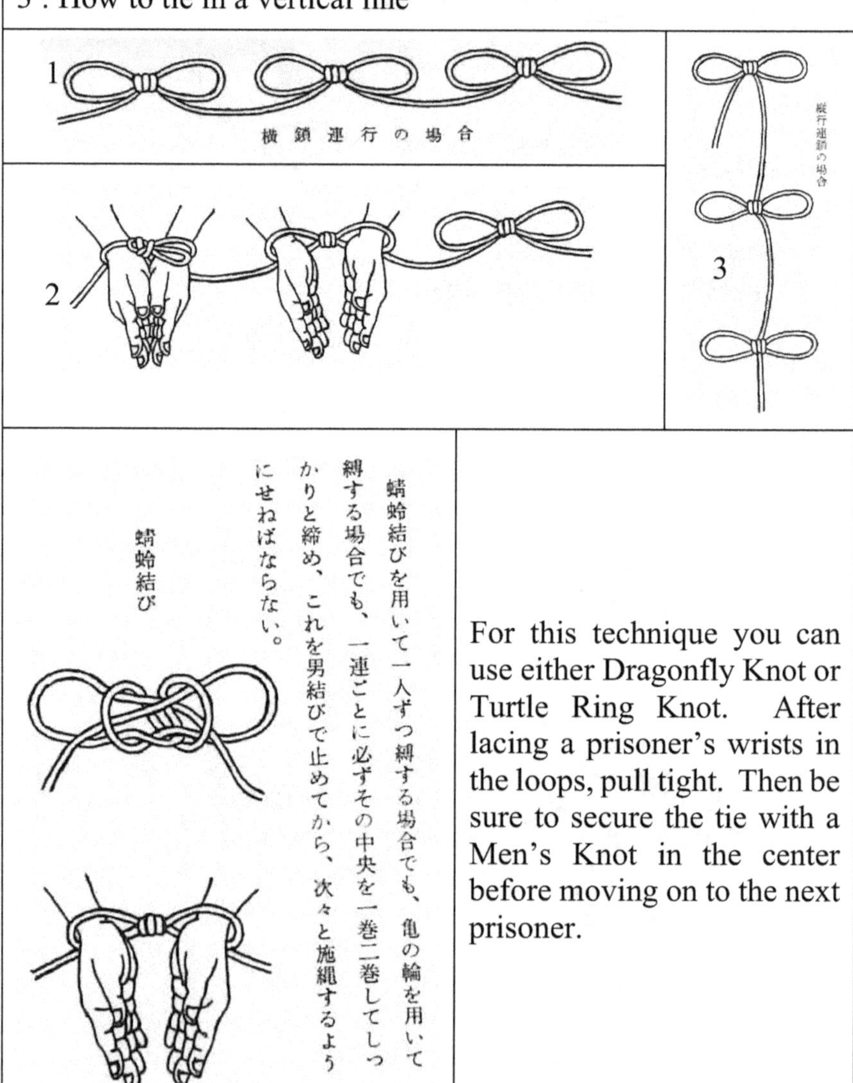

蜻蛉結びを用いて一人ずつ縛する場合でも、亀の輪を用いて縛する場合でも、一連ごとに必ずその中央を一巻二巻してしっかりと締め、これを男結びで止めてから、次々と施縄するようにせねばならない。

For this technique you can use either Dragonfly Knot or Turtle Ring Knot. After lacing a prisoner's wrists in the loops, pull tight. Then be sure to secure the tie with a Men's Knot in the center before moving on to the next prisoner.

Torikata: How to Arrest

Kame no Wa: How to tie Chain Tie with Turtle Ring Knot

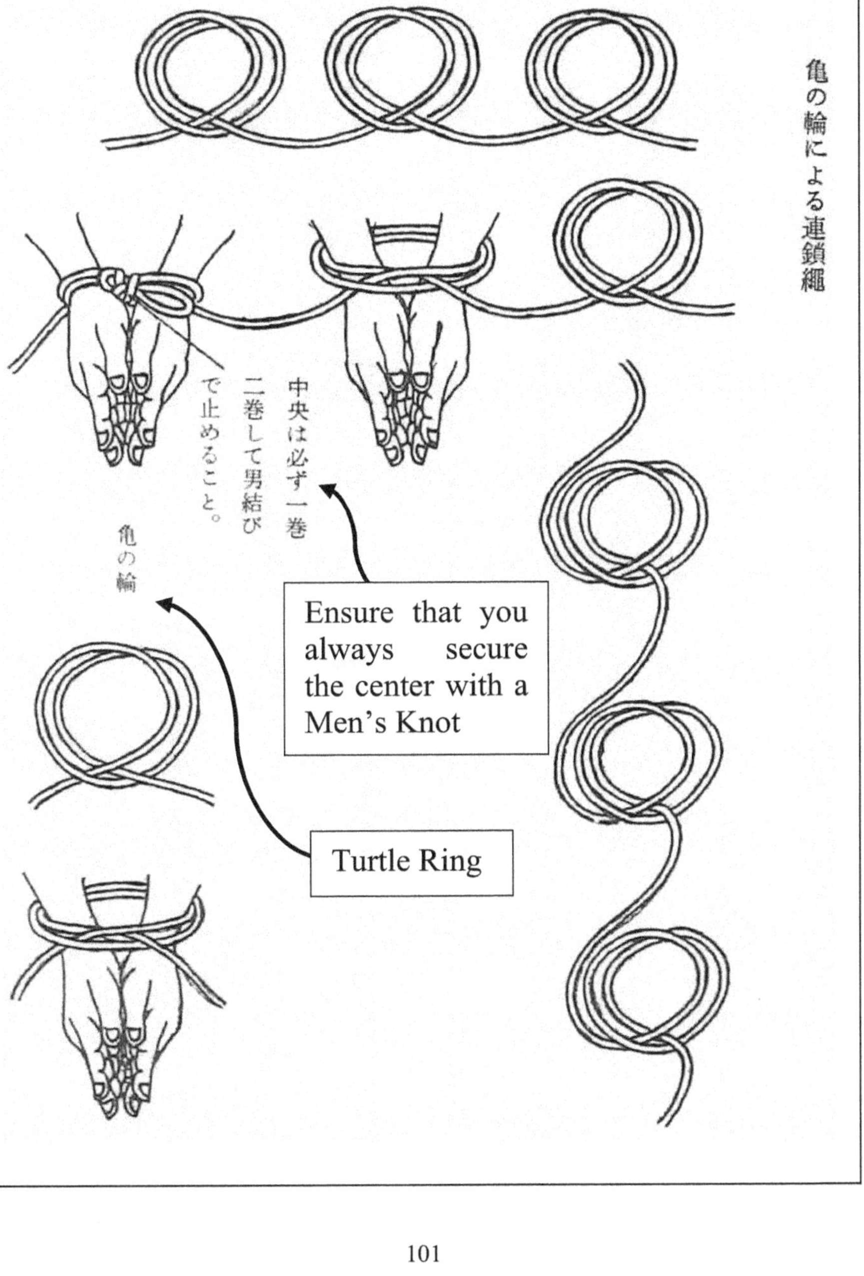

亀の輪による連鎖縄

中央は必ず一巻二巻して男結びで止めること。

亀の輪

Ensure that you always secure the center with a Men's Knot

Turtle Ring

Torikata: How to Arrest

Te no Tome Kata: How to Secure the Hands

手の留め方

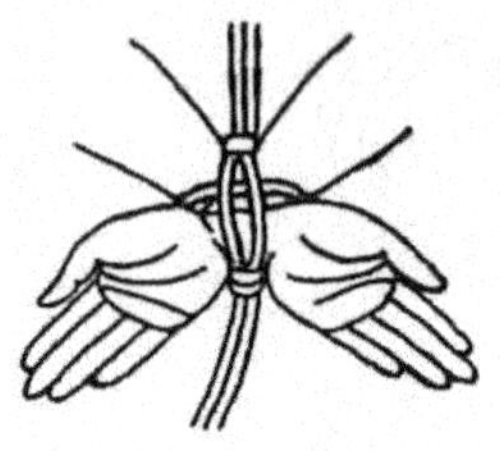

十字態に留る。

Tied off with hands crossed

Tied off with hands holding onto arms

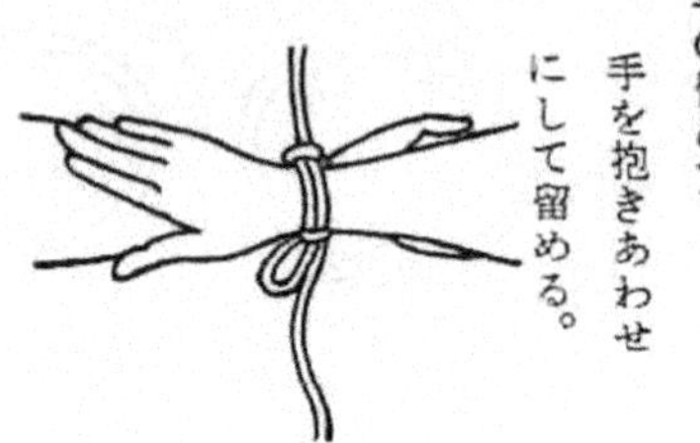

手を抱きあわせにして留める。

Tied off with hands overlapping

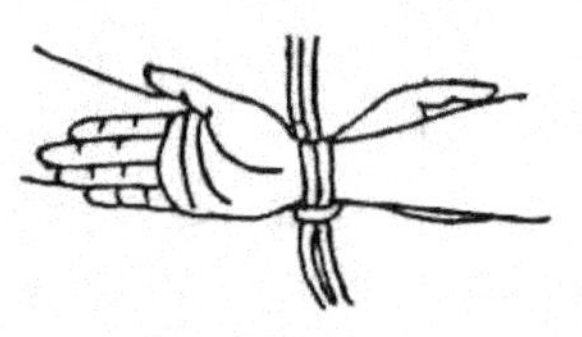

前後に重ねて留める。

Tied off with hands one on top of the other

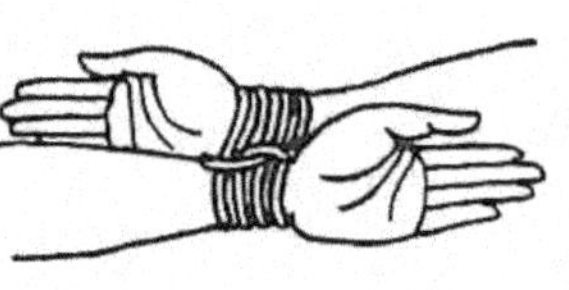

上下に重ねて留める。

Tied off with hands as if joined in prayer

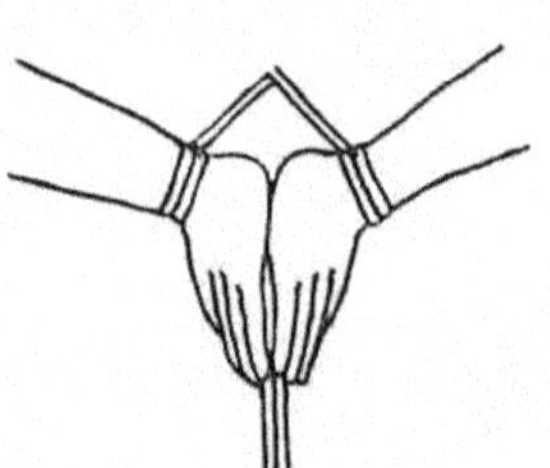

合掌体にして留める。

Hon Nawa

本縄

Ohso Yoh ・ Goso Yoh

押送用・護送用

Kakeyoh

掛様

Using Main Tie To Transport Prisoners Under Guard

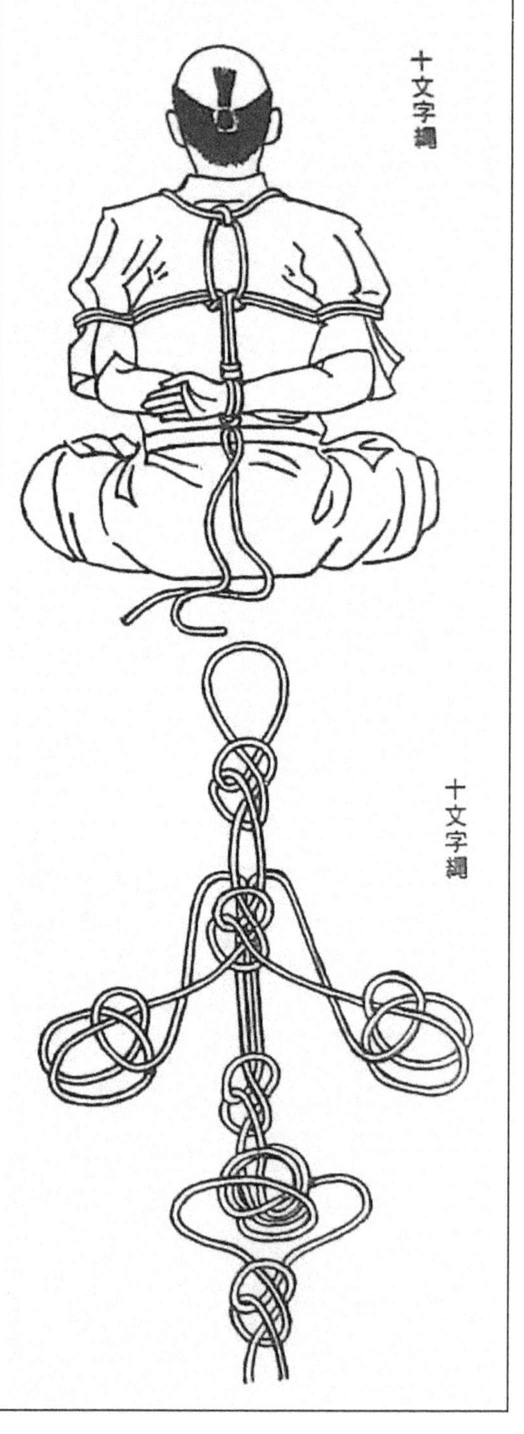

十文字縄これは雑人に掛ける縄なり。七尋の縄を折半し真中を首に掛け、垣根結びに結び、その縄を引き揃え背の中程にて結ぶ。それよりその縄を左右一筋ずつに分け、先ず一筋で左上腕を鎌がくしに縛り、次に右一筋で右上腕を同じく鎌がくしに縛り、左右両縄端を一束にして上の輪に背に面した方より通し、帯の上にて一束にして乳をつくり、乳の方を下にして縄の端を小手の上にして手の甲と節との間に回し、乳を引き締め、夫より縄を一筋ずつに分け、一筋を左右合せた手の間に引回して下の方にて垣根結びに結ぶ。

Using Main Tie to Transport Prisoners Under Guard 17 Techniques

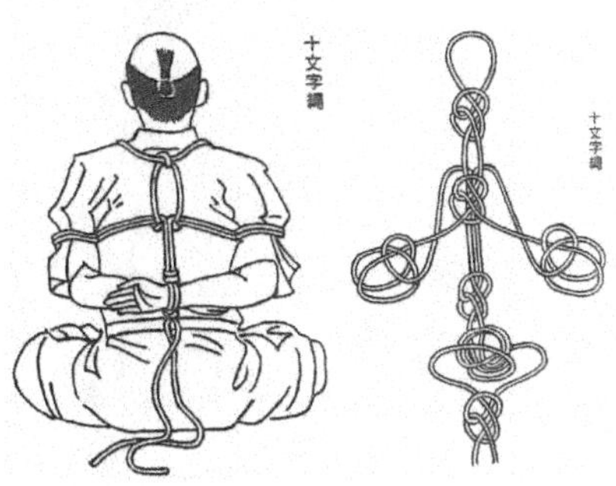

十文字縄 Jumonji Nawa Cross Shaped Tie

Note: The first Kanji means "ten" which is cross shaped.

This method is used on Zatsujin, people with no particular rank. First, select a rope 七尋 Shichi Jin, 7 Lengths, 7 arm spans approximately 180cm X 7 = 12 meters. Fold it in half and loop the center around the enemy's neck. Pull the two ends of the rope down until they are even and tie a Kakine Musubi, Fence Knot, in the center of the back. Take one strand in each hand and, with the left end, wrap the left upper arm with Kama Gakushi, Hidden Sickle Knot. Next, take the right end and tie the right upper arm with the same Hidden Sickle Knot.

After that, bring the ropes together again and pass them through the ring (the initial loop around the neck) on the back. You should pass the ropes through the top of the loop not under. Then tie another knot just above the belt and make a loop. The loop should be hanging down. Take one of the prisoner's wrists and place it in the loop. Twist the rope on top of his wrist and slip the other hand through. Pull the loop tight around the wrists. Split the rope in two and wrap one clockwise and one counterclockwise around the point the two wrists meet. Finally, bring the two ends together below and tie a Fence knot at the bottom.

Translator's Note: Fujita Seiko uses the Kanji 束 for "knot" and 乳 for "loop."

上縄　これも雑人に掛ける縄なり。この縄の形上の字に似たる故に上縄と名付けられた。
七尋の縄を折半して真中になるところを咽にかけ垣根結びに結び、その縄を左右に分け、一筋を以て先ず左上腕を鎌がくしに縛り、次に右上腕を同じく鎌がくしに縛り、その縄端を互い違いにとって咽と腕との間の縄へ引き掛け腰にて一束にして引きくくりの乳を造り、この乳へ縄の端を一束にして通し、乳を引き締めて小手を留る。

上縄（かみなわ）

上縄（かみなわ）

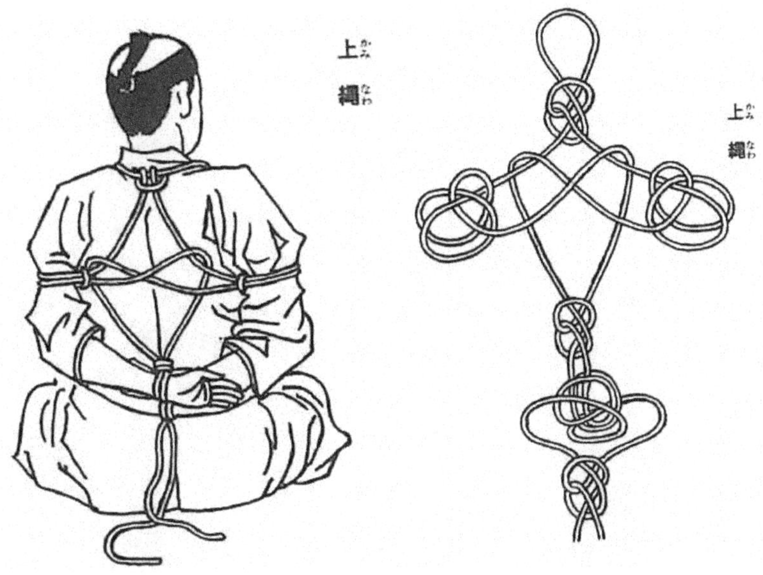

上繩 Kami Nawa: Upper Tie

Note: The Kanji 繩 is the older version of 縄. Both mean "rope."

This method is also used on Zatsujin, people with no particular rank.

First, select a rope 七尋 Shichi Jin, 7 arm-spans, 12 meters in length. Fold it in half and loop the center around the enemy's neck. Tie a Kakine Musubi, Fence Knot. Take one strand in each hand and, with the left end, wrap the left upper arm with Kama Gakushi, Hidden Sickle Knot. Next, take the right end and tie the right upper arm with the same Hidden Sickle Knot.

Take the two ends of the rope and cross them. Slip the right end over the rope going from the neck to the left arm and the left end over the rope going from the neck to the right arm. Bring the two ends together and make a knot at the waist. Then tie a Hiki-Kuguri, slip-knot loop. Bunch up the remaining rope and pass it through the loop and pull it tight around the wrists, securing them.

割菱縄

九尋の縄を二つに折り、その折った中央のところを左手に持って右手で二重のところを一束に持ち、被縛者の右脇下より後へ回し、左の肩に出し、内一筋を以て右脇の折り目の輪に通し、輪を胸部へ引き揚げ、残りの一筋は肩をはずし、被縛者の左脇下へ後の方より通し、その縄端を背中へ筋違いにかけて右肩へ執り、左折返しの輪へ通しこれも胸部へ引き揚げて左右を締め寄せる（後は襷を掛けた如くになる）。俗胸部に括り寄せて垣根結びにしてその縄を抜き通さず乳を造り、乳になしたる縄を一重巻いて締めれば乳ずれることなし、それより縄を左右に分け腕に掛け、左右の縄を束ね裡より外へ一束に通して小手を留る。

この縄は常は雑人に用うるが、旅押国渡にも用ゆ。食事の時小手を解き縄端を乳に結んで置く。

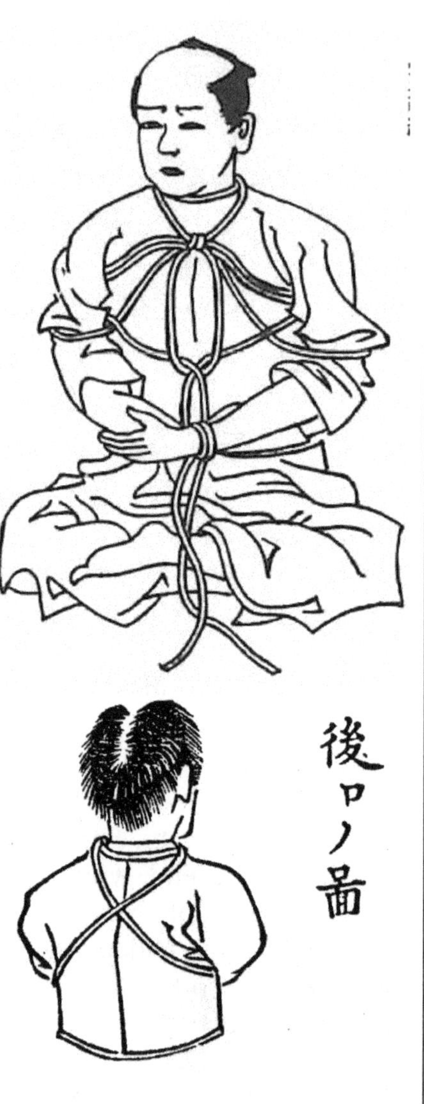

View from the back

割菱縄 Wari Hishi Nawa: Split Water Chestnut Tie

First fold the rope in half and hold the center bend with your left hand in front of your prisoner's chest. Take the two ends in your right hand and bring them under his right armpit across his back and over his left shoulder. Thread one end (around his neck and) through the center loop you are holding in front of his chest with your left hand. Pull the rope tight so the loop is firm against the chest.

Take the other strand you just brought over his left shoulder (and pass it through the loop on his chest,) under his left armpit and over his left shoulder. This will result in an X being formed across his back. Bring that cord (around his neck) and thread that end through the loop on his chest.

Bring the two ends together in front and pull the loop tight around his chest. Tie a Kakine Musubi, Fence Knot. The back should look like you tied a Tasuki sleeve cord. To ensure the rope does not slip off make a loop in front. Making a loop out of doubled rope and tying it off will prevent slippage.

Then separate the two strands and wrap each arm by passing the rope over the front of the arm, under the armpit and back towards you, finally passing through the loop on the chest. The last step is tying the hands.

This method is typically used on Zatsujin, people with no particular rank, but can also be used when transporting prisoners to other domains in Japan. At mealtimes the ties around the prisoner's hands can be removed and a loop made to attach to some fixture.

Translator's Note: The illustration showing the back is not from Fujita Seiko's book but from Inoguchi Matsunosuke's 1899 book *Illustrated Guide to the Secrets of Martial Arts.*

違菱縄 Chigai Hishi Nawa: Differing Water Chestnut Tie

This method is also used on Zatsujin, people with no particular rank.
First select a rope 7 arm-spans or 12 meters in length. Fold it in half and loop the center around the enemy's neck. Tie a Fence Knot and make a loop under that. Divide the ropes left and right. Wrap first the left arm and then the right arm and bring the ends back through the loop on the back. Make a knot at the waist and tie a slipknot. Tighten the slipknot around the hands to secure.

違菱縄　この縄も雑人に掛ける縄なり。
七尋の縄を折半し、その中央部を咽にかけて垣根結びに結び乳を造り、縄を左右に分け、左右の順に腕を縛り引揃え乳に通し、腰にて一束に引括りの乳を造り小手を留る。

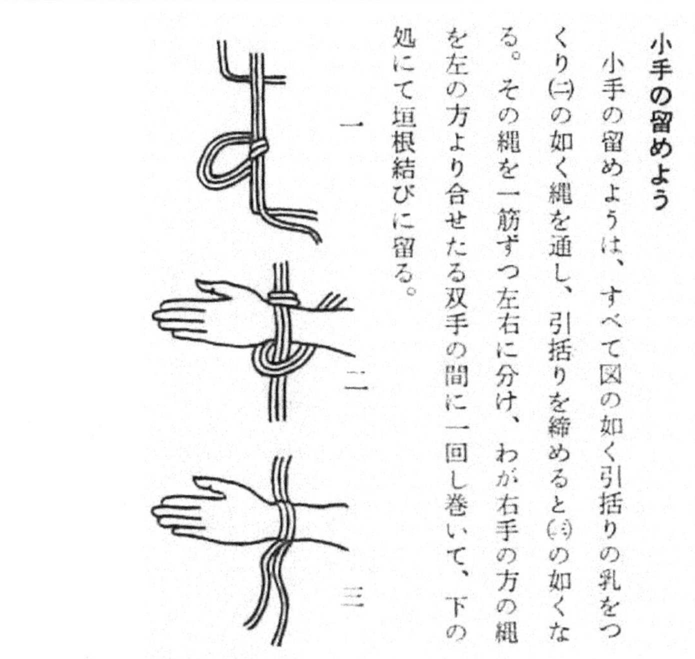

小手の留めよう

小手の留めようは、すべて図の如く引括りの乳をつくり(二)の如く縄を通し、引括りを締めると(三)の如くなる。その縄を一筋ずつ左右に分け、わが右手の方の縄を左の方より合せたる双手の間に一回し巻いて、下の処にて垣根結びに留る。

Kote no Tomeyo:
Securing the Hands

The illustrations show how to make a Hiki-kukuri, Slip-knot loop, to secure the hands.

After making (1) pass the hand through as shown in (2) and pull it tight so it looks like (3.)

Divide the rope into left and right strands and wrap each in opposite directions between the hands one revolution. Finally, tie a Fence Knot at the bottom to secure.

下廻縄この縄は剛力者に掛ける縄にて常人には余り用いないが、両足を縛って歩行を止むる縄であるから、時によって常人に用いてもさしっかえはなし。

七尋ないし九尋の縄を用う。掛け方は折半せる縄の真中を首に掛け、前にて垣根結びにして乳を造り、縄を左右に分けて左右の腕を鎌がくしに縛り、その縄を臍のところにて一束に結び、引括りの乳を造り、左右の足を組ませ、上なる縄を一束に乳に通し、組んだ両足を締め小手を留る。そしてその縄を股間を後へ引貫き、腰にて一束に結び、縄を分けて左右の脇の下より腹部に回し、また一束にして乳へ引き通し小手を留る。

下廻縄 Gekai Nawa: Downward Revolving Tie

This knot is reserved for tying up strong men and is typically not for average people. However, since it includes a tie to prevent walking, it is occasionally employed with average people.

Select a rope between 7 and 9 arm-spans, 12 ~ 16 meters. First fold the rope in half and loop the center around the neck of the enemy. Tie a Fence Knot in front and make a loop. Divide the ropes into left and right strands and tie a Kama Gakushi, Hidden Sickle Tie, on each arm. Bring the ends together and tie a knot above the navel. Then tie a Hiki Kukuri, Slipknot. Have the enemy cross his legs and place them in the Slipknot. Pull the end of the rope, which is facing up to tighten the Slip-Knot tight around the legs. After that tie off the wrists.

Pass the ends of the rope under the thighs and out behind him. Tie a knot there and pass one end under each armpit around to the stomach and through the loop on the front. Tie another knot there and make a Slipknot. Tighten the Slipknot over the hands to further secure them.

返し縄出家にかける縄なり。
掛け方は割菱と同じく脇の下より襷掛けにして括り乳を造り、縄を左右に分け、左腕右腕と鎌がくしにして左右の縄端を一束に乳の裡より外へ引き通し、帯の上にて小手を留る。出家は袈裟を外して縄を掛けるのが定法である。また衣も脱がせる方がよく、袈裟を脱すれば僧衣も共に脱するものと心得るべし。

鷹の羽返し縄これも出家に掛ける縄なり。返し縄の如く腕を縛りその縄端にて前腕を締り、その端を乳に通し小手を留るなり。

Kaeshi Nawa: Returning Rope

Taka no Hagaeshi Nawa: Hawk's Wing Returning Rope

返縄 Kaeshi Nawa: Returning Rope

This tie is used for Shukke, travelling monks.

This tie begins in the same way as Wari Hishi, Split Water Chestnut. Except it is done on the back instead of the chest. The rope passes underneath the armpits and forms an X on the chest like when tying the sleeves of the Kimono up with the Tasuki cord. Tie a Slipknot in the center of the back and separate the two strands. With one strand tie off the right arm with a Kama Gakushi, Hidden Sickle Knot, and then do the left arm with the other strand.

Pass the strands of rope through the loop from behind and out. Secure the hands just above the belt.

It is standard practice to remove the Kesa, or monk's vestment worn by travelling monks, before tying the Hojo. In addition, the cloak worn by the monk is typically also removed. Remember that when the monk's vestment is removed, the monk's robe is typically also removed.

鷹の羽返縄 Taka no Hagaeshi Nawa: Hawk's Wing Returning Rope

This tie is also used with Shukke, travelling monks. It is basically the same as Kaeshi Nawa, except it is done on the front. After the arms are tied, take the ends of rope and tie the forearms. Then make a Slipknot and secure the hands.

The Emperor Go-Mizunoo 後水尾天皇 (1596-1680) wearing a Kesa. The Japanese Kesa were worn over the left shoulder. Some Kesa were robes unto themselves.

注連縄この縄は社人に掛ける縄なり。その形が注連に似ているので名付けられたもの。掛け方は返し縄の如く、まず襷に掛けそれを襟に括り寄せて、垣根結びにして乳を造り、次に縄を左右に分け、一筋にて左の前腕を鎌がくしに締り、また一筋にて右の前腕を鎌がくしに縛る。そしてその前腕の縄を似て左の上腕を鎌がくしに締り、また右も同じく鎌がくしに縛り、この縄を一束にして乳に引き通し、小手を留める。

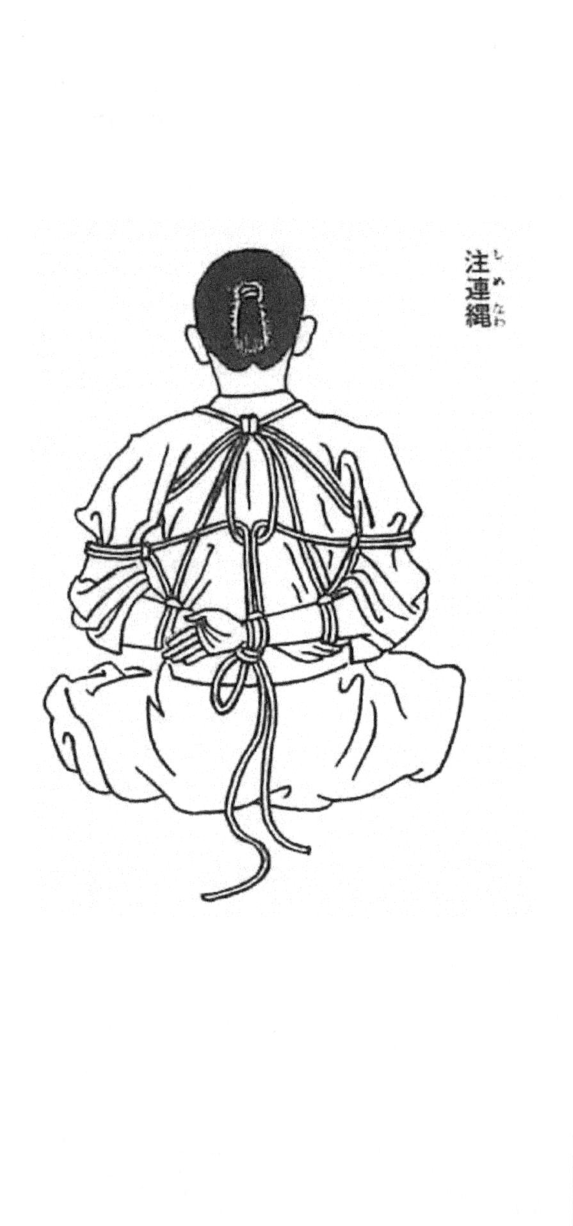

注連縄 Shime Nawa: Binding Rope

This tie is used to secure priests from Shinto shrines. It is called Shime becaue it resembles the rope hanging in front of Shinto shrines.

This tie starts out like Kaeshi Nawa. Make the ropes cross the back in an X like the Tasuki and fasten below the collar in Fence Knot. Then tie a loop below that.. Separate the two strands of rope. With one strand tie Hidden Sickle Knot around the left forearm, and the with the other tie a Hidden Sickle Knot around the right forearm.

Next, take each end of rope and tie a Hidden Sickle Knot around the biceps of the right and left arms. Pass the two ends through the loop in the back and secure the hands.

Note: Pictures of Shime Nawa at Shinto Shrines

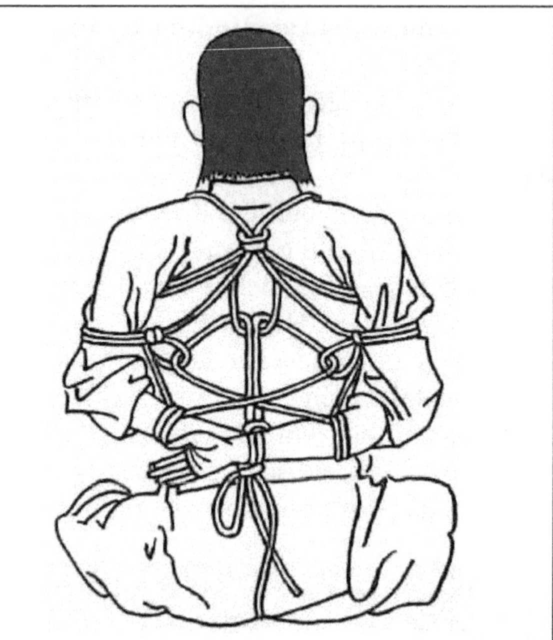

笈植縄 Oishoku Nawa:
Bamboo Backpack Tie

This tie is used to restrain Yamabushi, mountain ascetics.

This technique starts off like the previous one. Cross the ropes like the X formed when wearing the Tasuki cord. Make a loop in the center of the back and split the strands left and right. Tie of the upper am on each side with a Hidden Sickle Knot and add a loop onto each. Then pull the ropes down to tie another Hidden Sickle Knot around the left and right forearms.
Pass the left strand through the loop on the upper right arm and pass the right strand through the loop on the upper left arm. Bring the two strands together and pass them through the loop on the back, from back to front. Finally, secure the hands.

笈植縄これは山伏に掛ける縄なり。
掛け方は前の如く縄を襷に掛け乳を造り、縄を左右に分け、左右上腕を鎌がくしに縛り、左右共に乳を造って置いてその縄端にて左右の前腕を鎌がくしに縛り、左の縄端を右の上腕の乳に通し、右の縄端を左の上腕の乳に通しこれを中の乳へ一束に裡より外へ引き通して小手を留るなり

羽付縄

対決等の場合に掛ける縄で、小手を留めない。これは時に貴人の前に出し手を突き挨拶させねばならぬこともあり、また食事の場合自由に双手を動かしめたり、縄付のまま馬に乗せねばならぬ時、鞍の前輪に取り付かせるためである。縄の掛け方は前と同じく襷に掛け、肩に括り寄せて垣根結びに結び、それより左右の腕を鎌がくしに縛り、それを腰にて一束に結び左右へ分け前へ回し、打ち違えにより後口へ回し腰にて垣根結びになし、縄端を腰に引き通して置くなり。

羽付縄 Hazuke Nawa: Fixed Wing Tie

This tie is used for Taiketsu, Judges, or people in similar professions. The hands are not secured. This is because the person may have to be presented in front of people of high rank and have to perform greetings. In addition, they have to be given freedom of movement in order to take meals or ride a horse. In the case of the latter, the front loop can be attached to the saddle.

This tie is done the same way as the previous technique. The rope should cross in an X as if a Tasuki cord is tied across the back. Bring the two strands over the shoulders and tie a Fence Knot in the center of the back. Separate the strands and tie a Hidden Sickle Knot on each arm. Bring the strands together and tie a knot at waist level. Separate the strands and wrap them around the front of the prisoner. Cross the two strands in front of the abdomen and bring them behind again. Tie a Fence Knot in the small of the back. Push the remaining rope through the Obi, or belt.

Note:
対決 Taiketsu: Judge
In the Kamakura to Muromachi Era this was one way to resolve disputes. If two parties could not agree on a matter the went before Taiketsu Judge and each got to plead their case.

乳掛縄この縄は貴賤に寄らず婦人に掛ける縄なれば乳掛けと名付けたのである。七尋の縄を折半し、その中央より少し片々へよったところを採って前方より婦人の右腕を鎌がくしに縛り大振の乳を造り、次に長い方の縄を婦人のうしろへ一文字に回し、その縄で左腕を同じく鎌がくしに縛りこれにも乳を造り、次にその双方の縄端を脇の下よりうしろへ回し背中にて打ち違いにとり、肩より双方の乳へ通し一束に結び、三寸くらい下にて一束に引括りを造り小手を留る。

乳掛縄 Chigake Nawa: Breast Tie

This tie is used to secure married women of any status, high or low. For that reason it is called Breast Tie.

First select a rope 7 arm-spans 12 meters in length. Fold it in half but take hold of it slightly off center. Stand in front of the married woman and tie a Hidden Sickle Knot on her right arm and then make a large loop. Next, with the long end of the rope, wrap the rope around the back of the married woman so it makes a line like the Kanji for the number one 一. Tie another Hidden Sickle Knot around the left arm and make a loop.

Next, take the ends of each strand and pass them under her armpits, cross them behind her and bring them over her shoulders. After that, thread one of the strands through each of the loops on her arms, gather the two strands together and tie a knot in the middle of her chest. About 9 centimeters below that knot tie a Hiki Kukuri, Slipknot and secure her hands with that.

足固縄この縄は船中に用うる掛け方で、また剛力者にも用いた。被縛者のうしろより彼の左上腕を鎌がくしに縛り、長い方を咽にかけそれを右へ回し、右上腕を鎌がくしに縛り被縛者の前へ回り右の一筋を採って左の股を鎌がくしに縛り、また左の一筋をとって右の股を同じく縛り、次に足を立てさせ締付け双方の縄を前にて打ち違えに採りうしろへ回し、そのまま咽へ掛かりたる縄へ通し背中にて一束に結び、その端に一束の引括りを造り小手を留るなり。斯の如く縛る時は下廻縄と同じく立つ事も叶わざれば剛力者に用うるも大いによし。

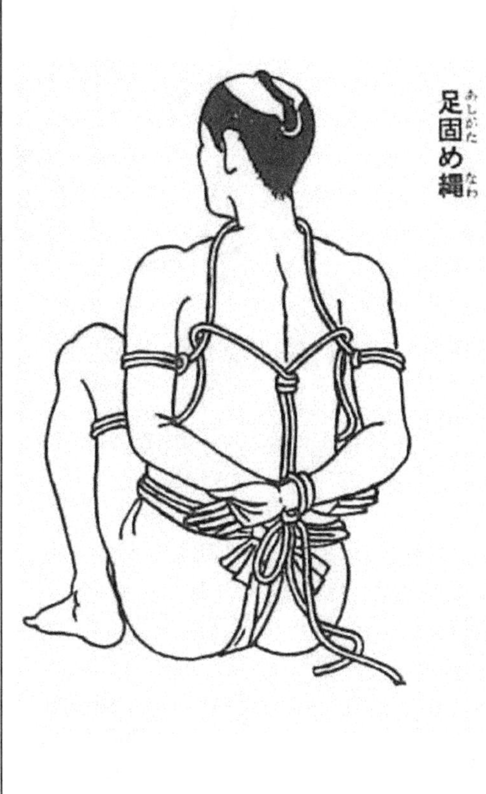

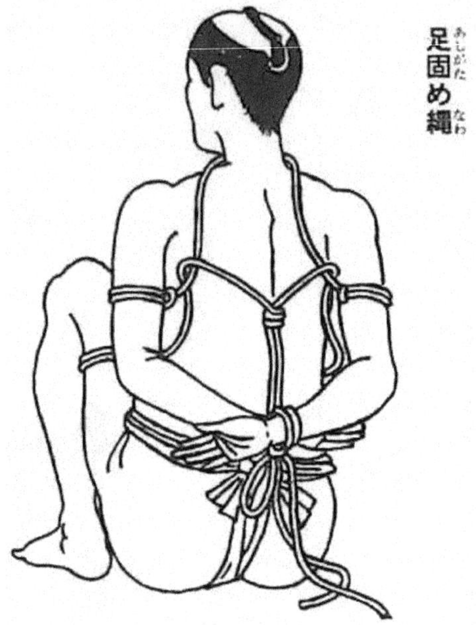

足固縄 Ashi Gatame Nawa: Leg Lock Tie

This tie is used when transporting a prisoner by boat or if the prisoner is strong and violent.

Standing behind the prisoner, tie a Hidden Sickle knot around his left upper arm. Loop the long end of the rope around his throat, wrapping from left to right. Tie a Hidden Sickle knot on his right upper arm.

Next, stand in front of the prisoner and take the right strand and tie a Hidden Sickle Knot around the left thigh. After that take the left strand and tie a Hidden Sickle Knot around the right thigh. After that, bend the legs up (Note: This must mean the legs were flat on the ground to tie the knots.) Take the ends of the two strands of rope and cross them over each other before wrapping them around the prisoner's back and passing the ends through the rope going around his neck. Make a knot in the center of his back and tie a Hiki Kukuri Slip Knot. Use this to secure the hands.

Just like with Gekai Nawa, even if a prisoner is very strong, they will find themselves unable to stand. It is very useful in these cases.

二重菱縄この縄は士に掛ける縄なり。
九尋の縄を折半し、その縄の真中を後襟に掛け、垣根結びにして乳を造り、縄を左右に分、右腕を鎌がくしに縛り、乳を造り、次に左腕を同じように縛って乳を造る。そしてその左右の縄を一束にして引括りを造り、左右の手を合せ鎌がくしに留め、それより縄を左右に分け、臍下より腰部のうしろへ回し、打ち違えにとり、前帯の上部で垣根結びにしてそれより左右の腕の乳に通し、縄端を一束に胸部の乳に通し、その乳に掛けてからげて置くなり。
この縄法を後に掛けるのも同じことである。途中連行の時は衣類の背に穴を明け縄を貫し、乳にからげて連行することもある。

二重菱縄 Niju Hishi Nawa:
Double Water Chestnut Rope
This tie is used to restrain a Samurai.

Select a rope 9 arm-spans, 16 meters, in length. Fold it in half and loop the middle of the rope around the back of the collar. Tie a Fence Knot and make a loop below that. Separate the two strands and tie a Sickle Knot around the Samurai's right bicep and make a loop. Next, do the same for the left bicep, tying a knot and making a loop.

After that bring the two strands together and tie a Hiki Kukuri Slip Knot. Secure the hands in the Slip Knot and tie it off with a Fence Knot. Separate the strands into left and right again and wrap them around the waist at the level of the navel. Cross the two strands behind his back and bring the two strands to the front and tie a Fence Knot around his belt.

Next, separate the strands into left and right and pass them through the loops on each bicep and in the center of the chest. The remaining rope can be secured to something as necessary.

This tie can be done on the back in the same manner as the front. When travelling on the road a hole can be cut in the back of the Samurai's clothing and the end of the rope passed thorough that. The rope can then be secured to something while travelling.

留り縄この縄は縄抜けの巧みな者にかける縄法である。九尋の縄を折半してその中央を咽に掛け、垣根結びにして乳を造り、それよりその一筋を以て左手首を鎌がくしに縛り、次に右の手首を同じく縛り、一束にして乳に引き通し、帯の引にて一束に引括り、被縛者に左右の指を組ませ、輪を掌の方にして縄を組合せたる指の節の間より掌の方へ回し、輪を通してそれを括り、その輪を左右に分け、指と手の甲の間へ上より割回し、下の方にて垣根結びに縛り、帯へ通してからげる。斯の如く前より掛けるもよし。

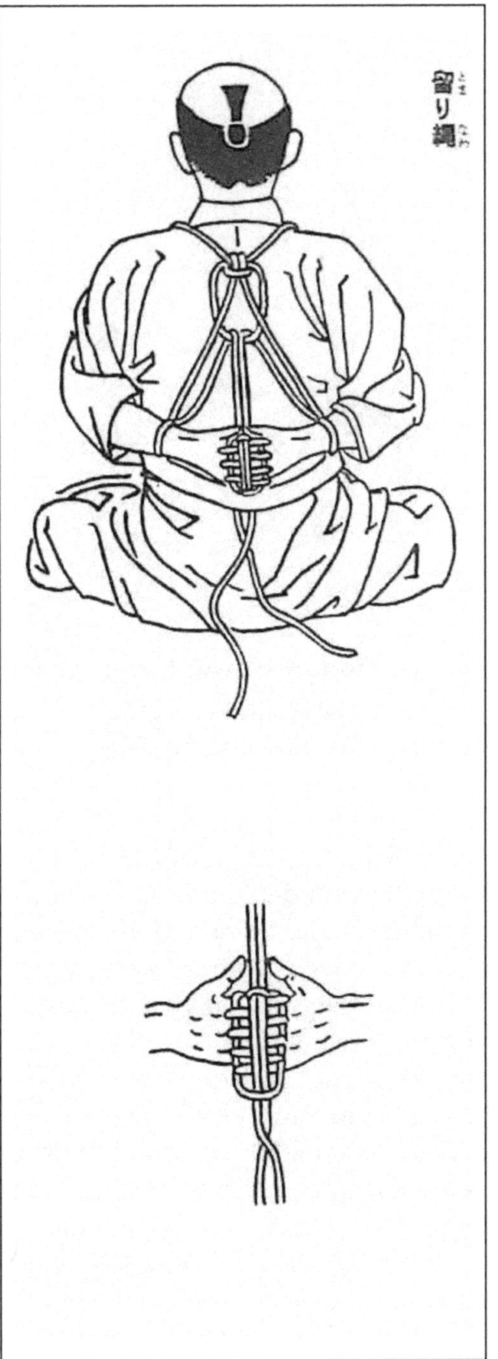

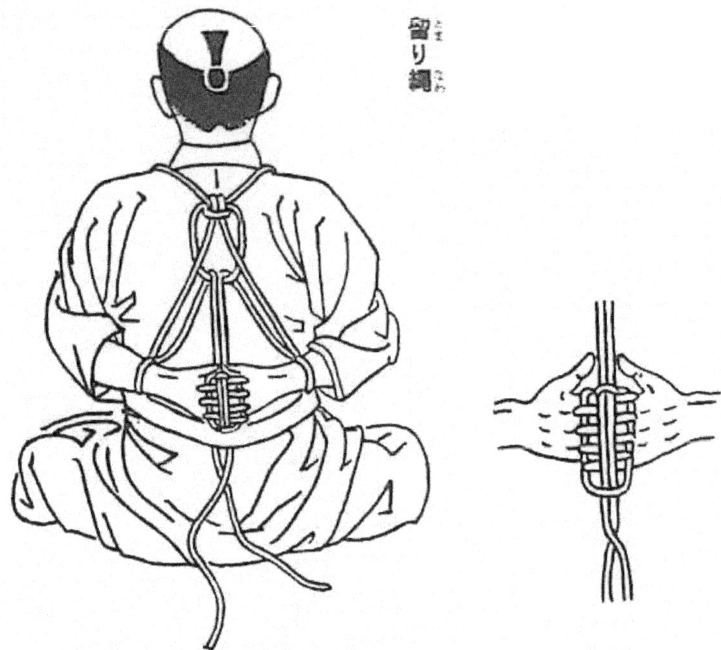

留り縄 Tomari Nawa: Stopping Rope

This tying method is used on prisoners who are adept at escaping despite being tied up.

Select a rope 9 arm-spans, 16 meters, in length and fold it in half. Place the center of the rope on the prisoner's throat and tie a Fence Knot and add a loop in the center of the back. Separate the two strands and tie one to the left wrist with a Fence Knot and one to the right wrist with the same knot.

After that pass the two strands through the loop on his back and bring the two strands together. Pass both strands through his belt and tie a knot followed by a Hiki Kukuri Slip Knot. Have the prisoner interlock his fingers and place the Slip Knot against his palms. Wrap the loop from below, over the back of the hands, but under the fingers. Pass the ends of the rope through the top of the loop, down out the bottom and pull the loop tight.

Separate the cords into left and right and wrap them in opposite directions between the fingers and the backs of the hands. Finally, tie a Fence Knot at the bottom and pass the end back through the Obi. The tie can be done the same way on the front.

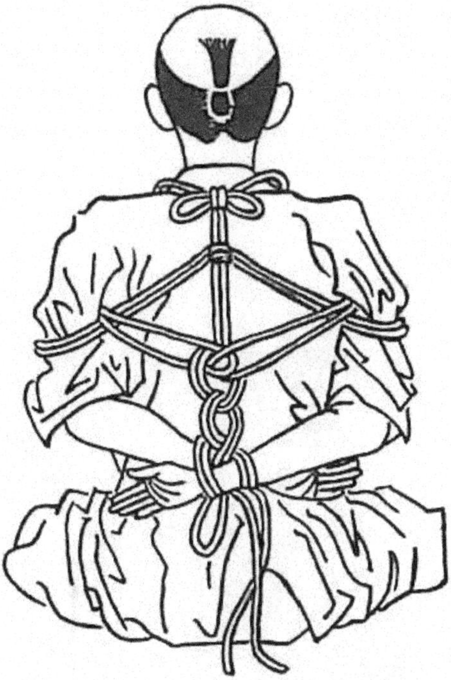

切縄この縄は首を斬る時に掛ける縄なり。

九尋の縄を折半し、一束のままにて咽を縛り、三寸ほど下ったところで一束に結び、その縄端を引き抜かず長目の大きな乳を造り、その乳を左右の上腕に掛け、脇の下より出し、その乳を背中にて一束に合せ、そしてその引き残した縄をこの乳に通し、その輪より鎖に組みその端にて小手を留るなり。首を斬る時は咽の縄を解いて首を斬らせ、小手を解き縄を引けば鎖は解けて、一度に縄が外れる。

なおこの切縄は、囚人を入檻させる時にも用うることがある。

切縄 Kiri Nawa: Executioner's Knot (Headsman's Knot)

This is the tie used on a person sentenced to decapitation by sword.

Select a rope 9 arm-spans, 16 meters, in length. First, fold it in half and wrap it around the prisoner's neck from the front. About 9 cm below that tie a knot. Next being certain not to use all the rope, tie a knot with two long loops. Wrap those long loops around both arms. From the back reach under the armpits and pull the loops towards you. Overlay the two loops and bring the remaining rope together before passing it through both loops. Tie the remaining rope in a Chain Knot and use the last portion to secure the wrists.

For the actual beheading by sword, the rope around the neck will be removed and the bared for cutting. This tic is also used when transferring prisoners to jail.

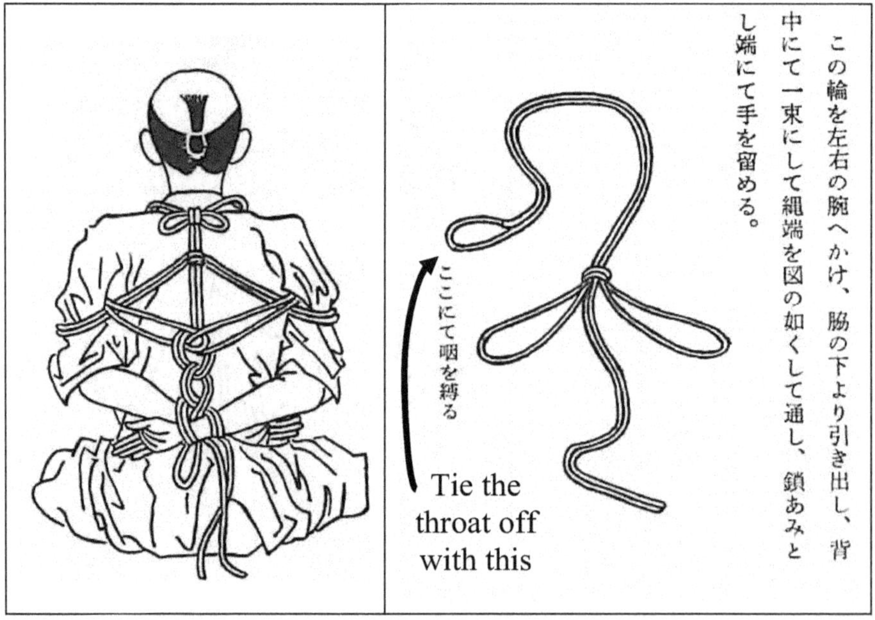

Text on right:

The two loops should go under the left and right armpits, wrap around the arms, pass under the armpits and then join in the center of the back. Bring them together and pass the remaining rope through them as shown in the illustration. Tie a chain knot to secure the hands.

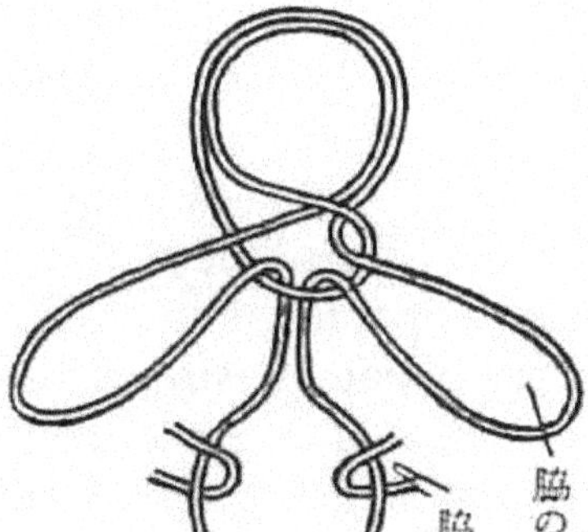

介縄この縄は囚人の受渡し追放放免等の場合に掛ける縄なり。九尋の縄を折半し中央輪の方を左手に採って咽に掛け、右手の持つ方を一束の輪に通し、一尺余り引き出し　(引抜くにはあらず　輪にして引き出すなり)それより輪になったところを左右に分け、上腕の上より前へ回し、その端を脇の下より後へ出し、この輪へ背に残した一筋の縄を通し、一重の鎖を三ずつ組み、端を一束に寄せ、背中へ二重の鎖を三つ四つ組み、その端にて小手を留るなり。

介縄（たすけなわ）

脇の下へ回す輪

脇の下より出したところ

これを片々ずつ鎖二、三を組み、また真中にて一束にして鎖三、四を組み、その端で小手を留める。

介縄 Kai Nawa: Transfer Tie

This tie is used when transferring a prisoner or for different degrees of banishment.

Select a rope 9 arm-spans, 16 meters, in length. Stand behind the prisoner. Fold the rope in half and hold it with your left hand. With your right-hand wrap this around the prisoner's neck and pull about 30 cm through the loop in your left hand. Be sure not to pull it all the way through. Divide the strands of the section you have pulled through the loop into left and right.

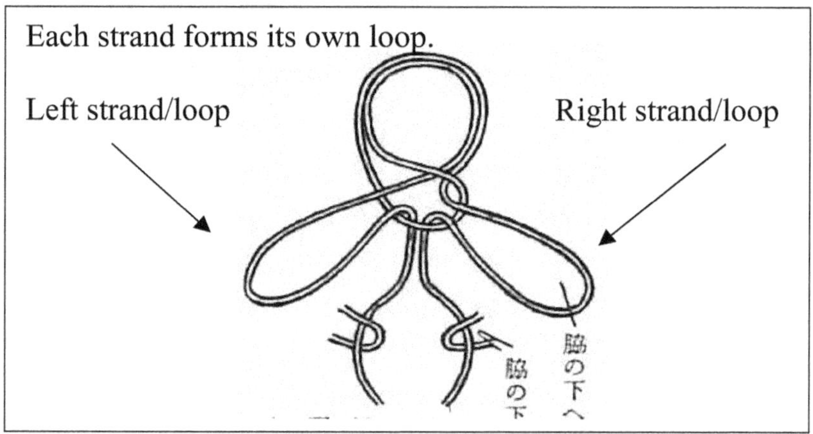

Slide the left strand over the back of the left arm and the right strand over the back of the right arm. Reaching under each armpit, pull the ends of the loops out.

Take the remaining strands of rope and pass one through each of the loops extending out of the back. Tie a doubled Chain Knot three times on each side. Bring the strands together and tie a knot. Then tie a double Chain Knot three or four times before securing the hands. When untying you only need to untie the hands and the rest of the rope will fall apart naturally.

People that are planning to escape after being tied always flex their body as the rope is being tied. Once they are tied up, they relax their body and the rope goes soft in places. All they need to do is find one spot that is loose enough to free one part of the body. Once a part of the body is free, the rest of the tie becomes easy to escape. When tying up a prisoner it is important to remember this.

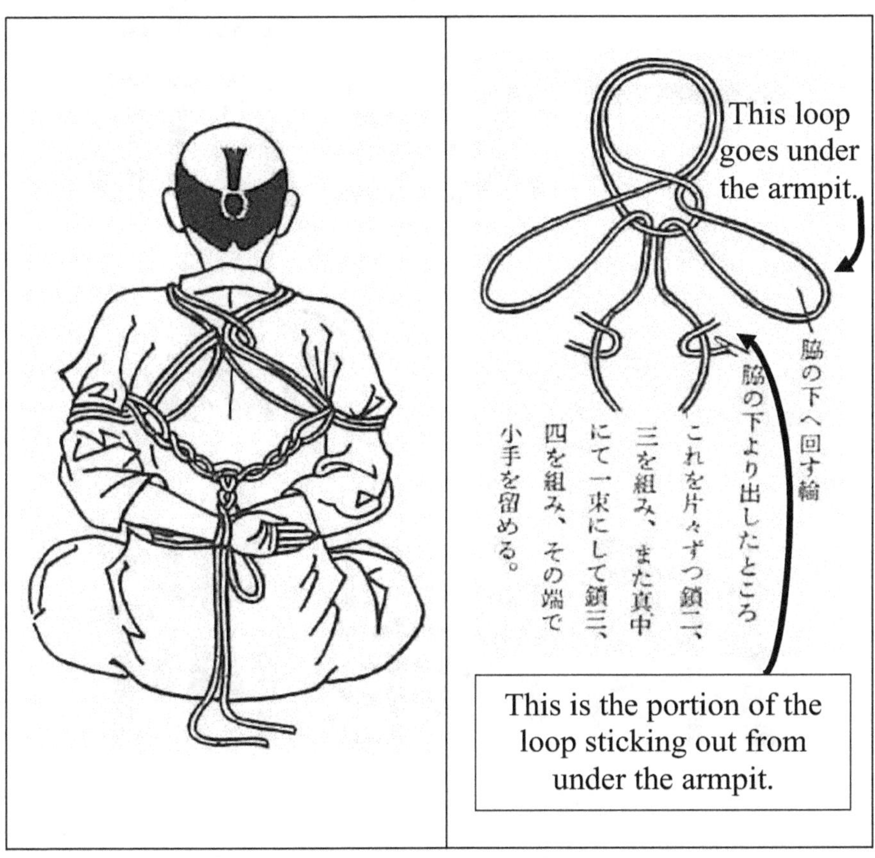

Text on bottom:
Each side should be tied in a Chain Knot two or three ties. Where they join together, tie a knot, and then tie another three or four Chain Knots before finally securing the hands.

Ittatsu Ryu

一達流

Hojo Jutsu

捕縄術

Bondage Techniques of the Ittatsu School

一達流

一文字 早
翅付 早
一重菱
真翅附
矢筈
角違
真蜻蛉
八方搦
櫓菱

菱 早
菱
十文字
馬上翅附
蜻蛉
真二重菱
真亀甲
櫓菱

十文字 早
十文字
二重菱
亀甲
揚巻
真翅附
胸割一重菱
切縄

Techniques of the 一達流捕縄術 Ittatsu School of Bondage Note: These techniques contain no descriptions.		
Kanji	Reading	English
早縄一文字	Haya Nawa Ichimonji	One Line Fast Tie
早縄菱	Haya Nawa Hishi	Water Chestnut Fast Tie
早縄十文字	Haya Nawa Jumonji	Cross Shaped Fast Tie
早縄翅付	Hagai Tsuke	Arm Lock
菱	Hishi	Water Chestnut
十文字	Jumonji	Cross Shaped
一重菱	Hito-e Hishi	Single Water Chestnut
十文字	Jumonji	Cross Shaped
真二重菱	Shin Futa-e Hishi	Doubled Water Chestnut
真翅附	Shin Hagai Tsuke	True Arm Lock
馬上翅附	Bajo Bagai Tsuke	Arm Lock on Horseback
亀甲	Kiko	Turtle Shell
矢筈	Ya Hazu	Nock of the Arrow
蜻蛉	Tombo	Dragonfly
揚巻	Age maki	Rising Wrap
角違	Sumi Chigai	Differing Corners
真二重菱	Shin Futa-e Hishi	Doubled Water Chestnut
真翅附	Shin Hagai Tsuke	True Arm Lock
真蜻蛉	Shin Tombo	True Dragonfly
真亀甲	Shin Kiko	True Turtle Shell
胸割一重菱	Munawari Hitoe Hishi	Split Chest Single Water Chestnut
八方搦	Happo Karami	Eight Way Wrap
櫓菱	Yagura Hishi	Watchtower Water Chestnut
切縄	Kiri Nawa	Executioner's Knot
早縄菱	Hishi Haya Nawa	Water Chestnut Fast Tie
櫓菱	Yagura Hishi	True Watchtower

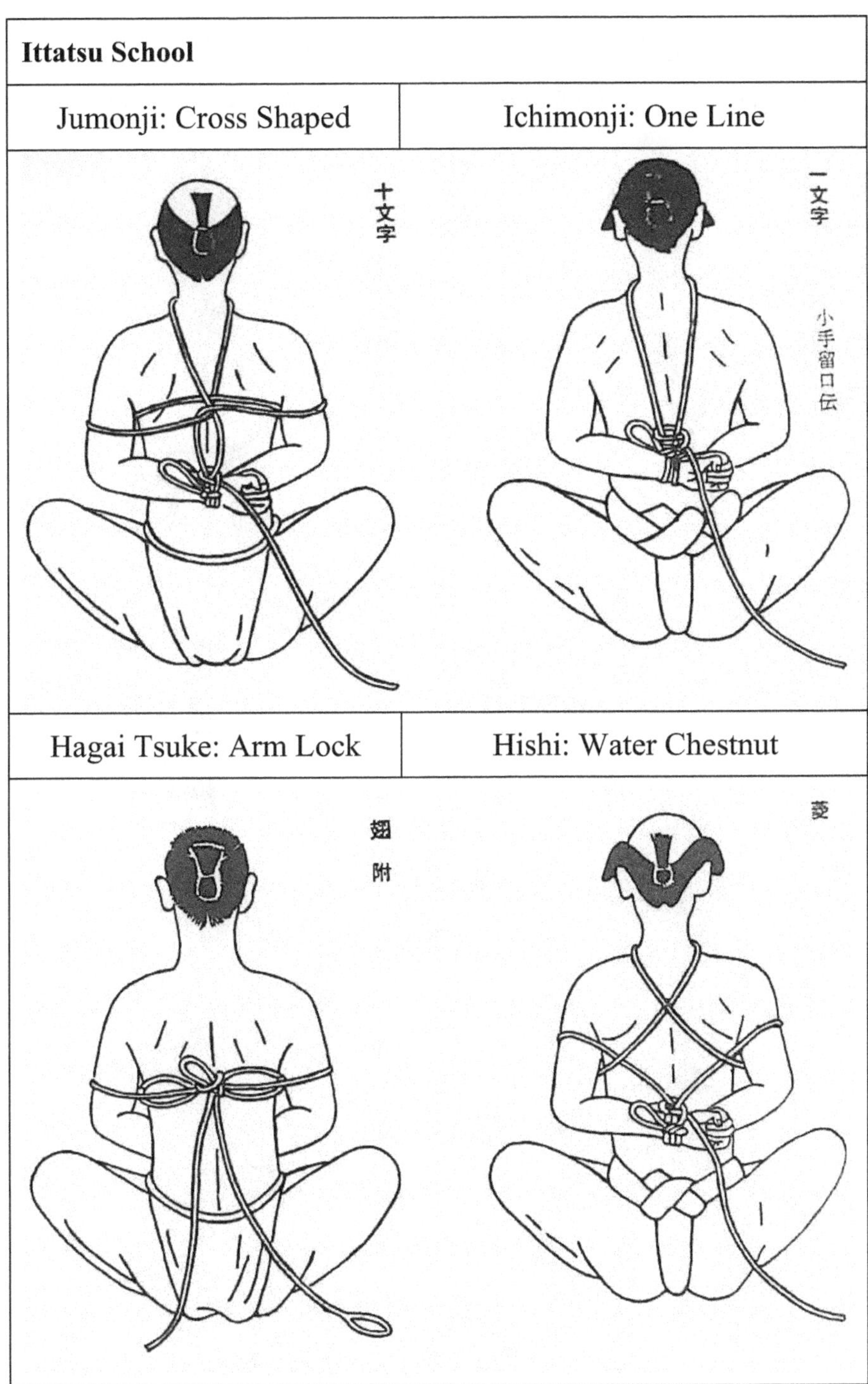
Ittatsu School
Jumonji: Cross Shaped
Ichimonji: One Line
十文字
一文字
小手留口伝
Hagai Tsuke: Arm Lock
Hishi: Water Chestnut
翅附
菱

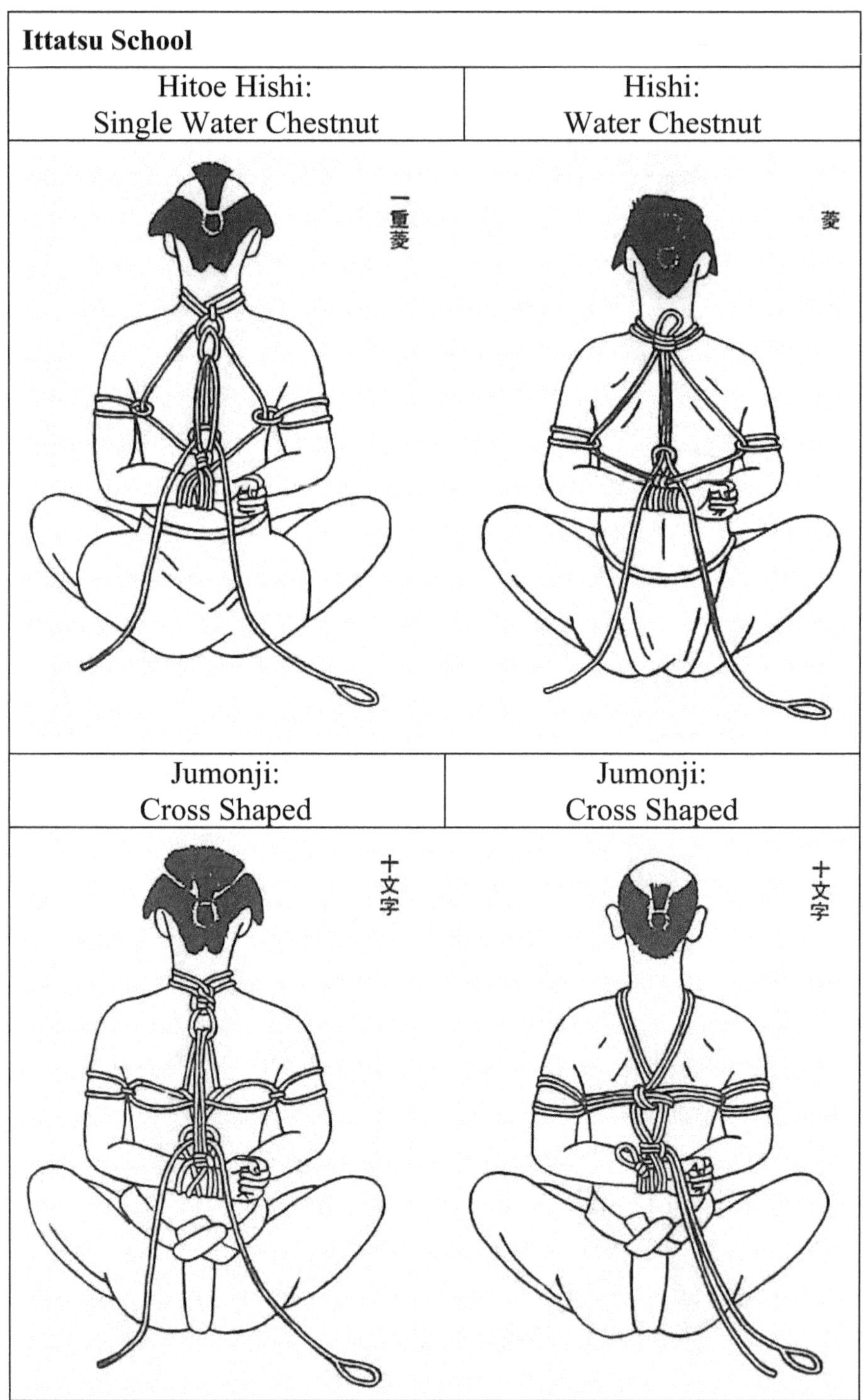
Ittatsu School
Hitoe Hishi:
Single Water Chestnut
一重菱
Hishi:
Water Chestnut
菱
Jumonji:
Cross Shaped
十文字
Jumonji:
Cross Shaped
十文字

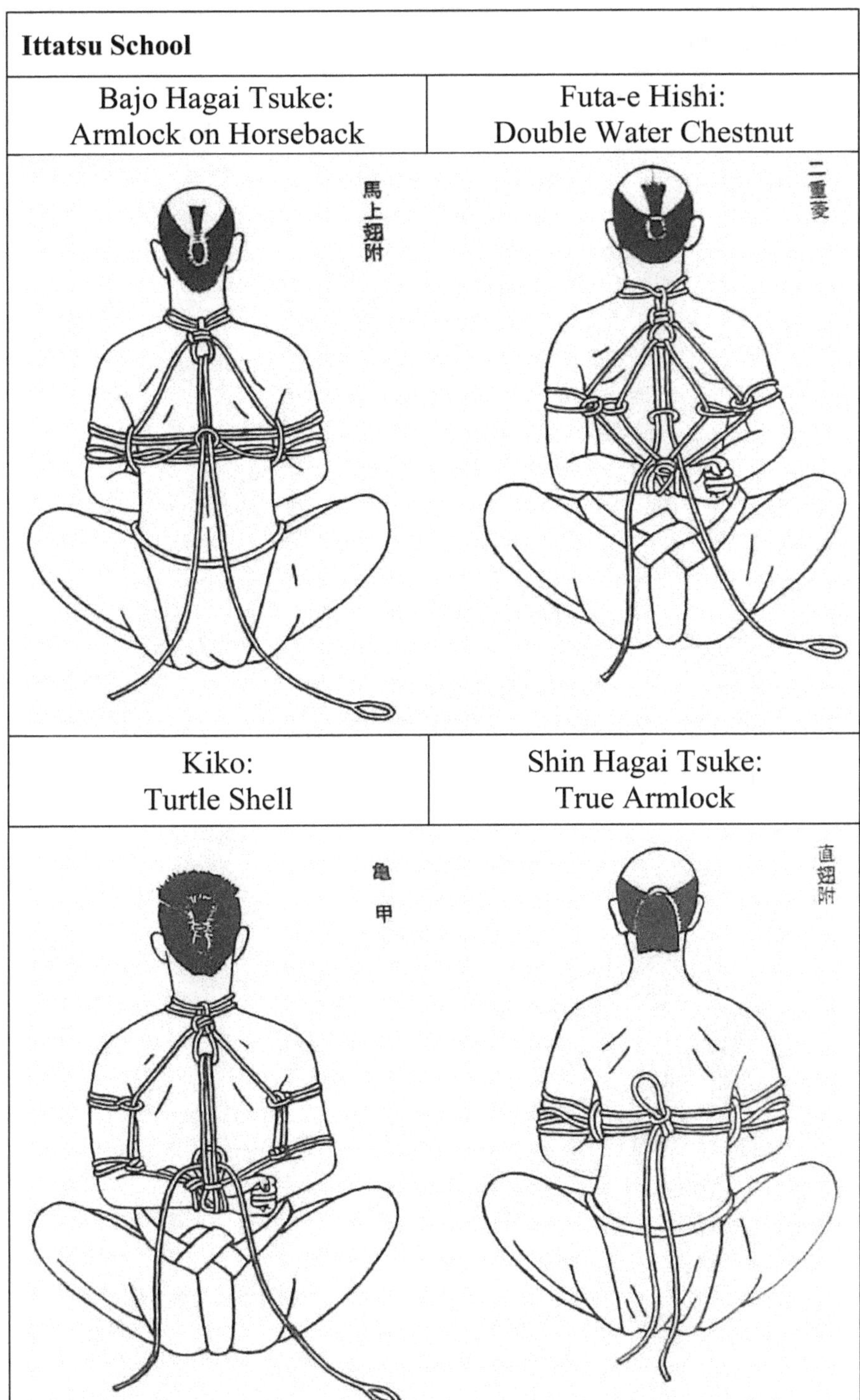

Ittatsu School

Bajo Hagai Tsuke: Armlock on Horseback	Futa-e Hishi: Double Water Chestnut
Kiko: Turtle Shell	Shin Hagai Tsuke: True Armlock

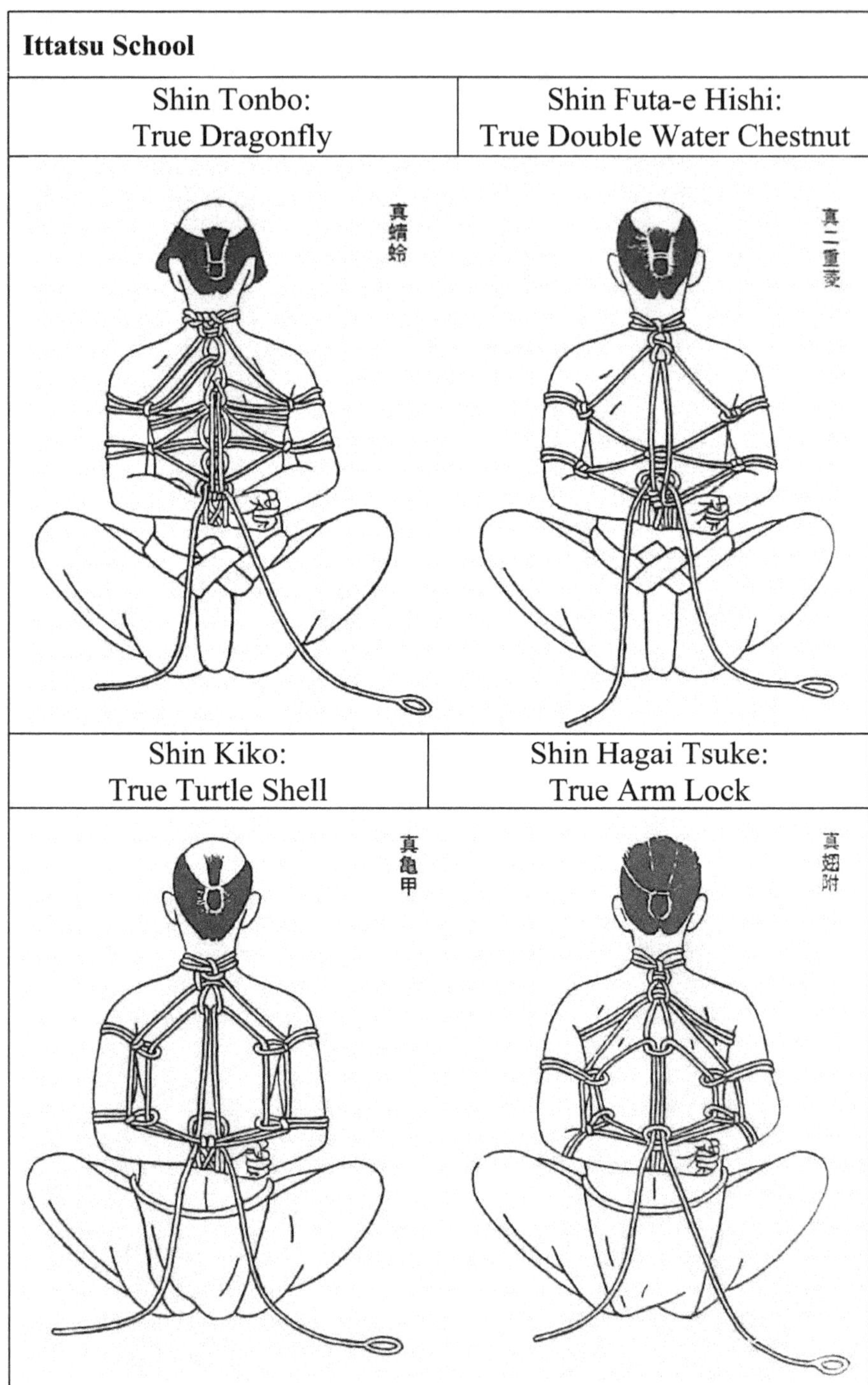
Ittatsu School
Shin Tonbo:
True Dragonfly
Shin Futa-e Hishi:
True Double Water Chestnut
真蜻蛉
真二重菱
Shin Kiko:
True Turtle Shell
Shin Hagai Tsuke:
True Arm Lock
真亀甲
真翅附

Ittatsu School	
Happo Karami: Eight Way Wrap	Mune Wari Hito-e Hishi: Split Chest Single Water Chestnut
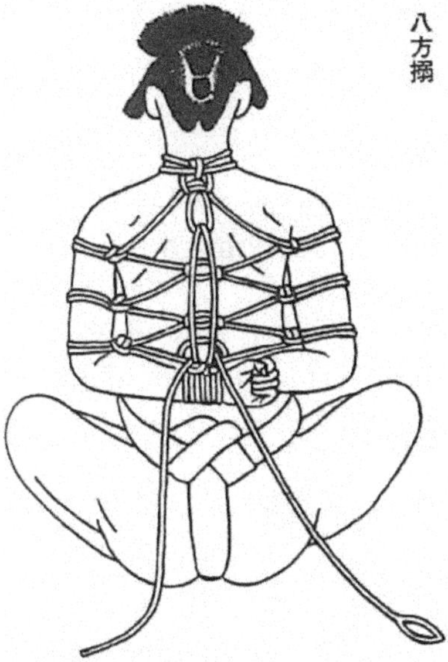 八方搦	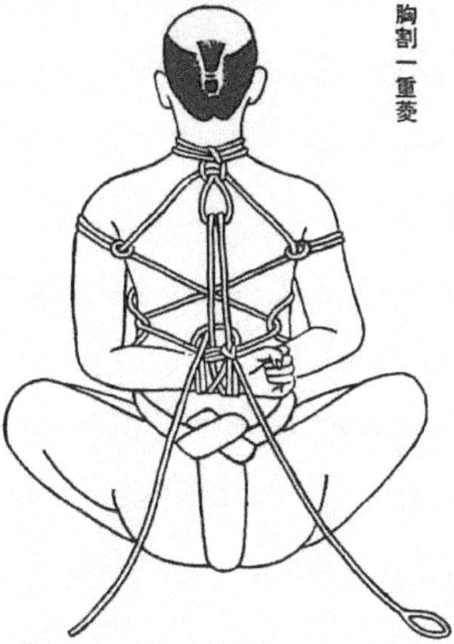 胸割一重菱
Yagura Hishi: Watch Tower Water Chestnut	Mune Wari Hito-e Hishi: Split Chest Single Water Chestnut (Front view of the above)
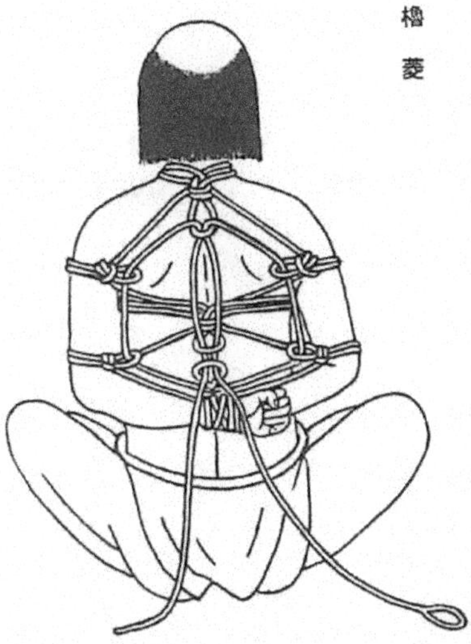 櫓菱	 胸割一重菱

Ittatsu School	
 切縄	Kiri Nawa: Executioner's Knot Note: This is probably a tie used when beheading a prisoner.
 櫓菱	Yagura Hishi: Watchtower Water Chestnut

Ho-en Ryu

方圓流

Hojo Jutsu

捕縄術

Bondage Techniques of the Ho-en School

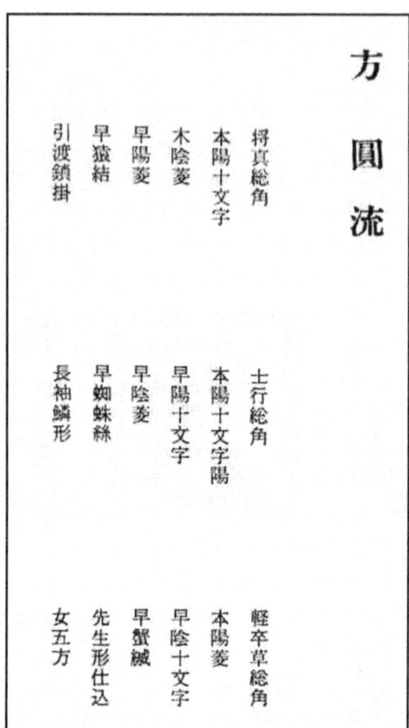

方圓流

将真総角
本陽十文字
木陰菱
早陽菱
早猿結
引渡鎖掛

士行総角
本陽十文字陽
早陽十文字
早陰菱
早蜘蛛絲
長袖鱗形

軽卒草総角
本陽菱
早陰十文字
早蟹縅
先生形仕込
女五方

Translator's Note:

There are no descriptions of these Ho-en Ryu techniques and no readings for the Kanji given. The readings in the following list are my best guess, as are the English meanings. The words Yin and Yang appear frequently since the duality was very important in martial arts. Two sides of the same coin.

Inyo is the Japanese word for Yin and Yang, so many of the following techniques will have In (Yin) or Yo (Yang) in the name. Interestingly they Yo is usually first, reversing the traditional order.

The first three techniques follow the Shin-Ko-So, True-Proceed-Grass pattern as mentioned at the beginning of this book. Shin-Ko-So originated in China and described the three different ways to write the characters that make up the Chinese language. Formally – Something in Between – Artfully. Shin represents the formal, prescribed method. So represents the artful/free application of the technique while Ko, is between the two.

真 *Shin* Represents Hon Nawa

草 *So* Represents Haya Nawa

行 *Ko* Represents both Hon Nawa and Haya Nawa used together.

Kanji	Reading (approximate)	English (approximate)
将真総角	Sho-Shin Agemaki	General's Formal Hairstyle
士行総角	Shi Gyo Agemaki	Samurai's Artful Hairstyle
軽卒草総角	Keisotsu So Agemaki	Foot Soldier's Something in Between Hairstyle
本陽十文字	Hon-Yo Jumonji	Main Yang Cross Shaped
本陰十文字	Hon-In Jumonji	Main Yin Cross Shaped
本陽菱	Hon-yo Hishi	Main Yang Water Chestnut
本陰菱	Hon-in Hishi	Main Yin Water Chestnut
早陽十文字	Haya-Yo Jumonji	Fast Yang Cross Shaped
早陰十文字	Haya-In Jumonji	Fast Yin Cross Shaped
早陽菱	Haya-Yo Hishi	Fast Yang Water Chestnut
早陰菱	Haya-In Hishi	Fast Yin Water Chestnut
早蟹繊	Haya Kani Sen	Fast Slice of Crab
早猿結	Haya Saru Ketsu	Fast Monkey Knot
早蜘蛛絲	Haya Kumo Ito	Fast Spider's Web
先王形仕込	Sen-O Kata Shikomi	In the Shape of our Long Past Lord
引渡鎖掛	Hiki-Watashi Kusari Gake	Handing Over Chain Tie
長袖鱗形	Nagasode Uruo Gata	Long Sleeved Fish Scale Shape
女五方	Onna Goho	Five Different Women

<table>
<tr><td colspan="2">Ho-en School</td></tr>
<tr><td>Keisotsu So Agemaki: Foot Soldier's Something in Between Hairstyle</td><td> Sho-Shin Agemaki: General's Formal Hairstyle
Left: Illustration of Agemaki hairstyle</td></tr>
<tr><td>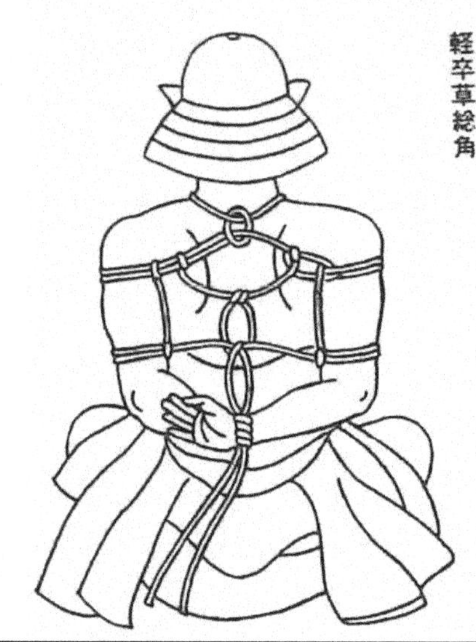</td><td>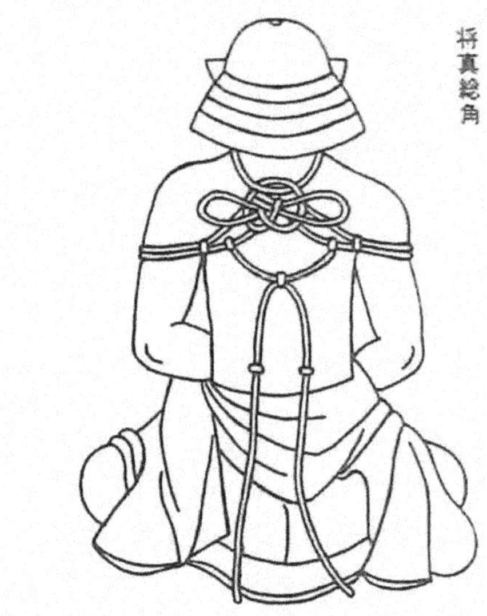</td></tr>
<tr><td>Hon-Yo Jumonji:
Main Yang Cross Shaped</td><td>Shi Gyo Agemaki: Samurai's
Proceeding Hairstyle</td></tr>
<tr><td>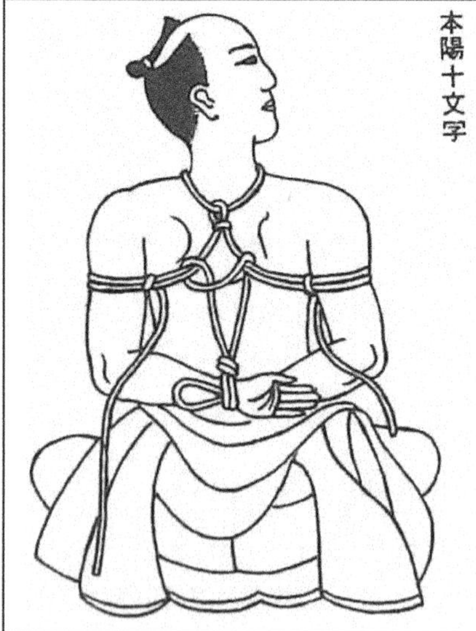</td><td>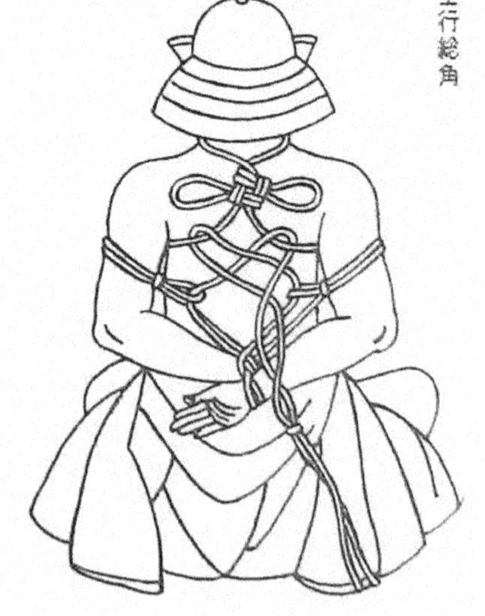</td></tr>
</table>

Ho-en School	
Hon-In Hishi: Main Yin Water Chestnut	Hon-In Jumonji: Main Yin Cross Shaped Note: The Kanji should be In 陰 since the last technique was Yo 陽
本陰菱	本陽十文字
Haya-Yo Jumonji: Fast Yin Cross Shape	Hon-Yo Hishi: Main Yang Water Chestnut
早陽十文字	本陽菱

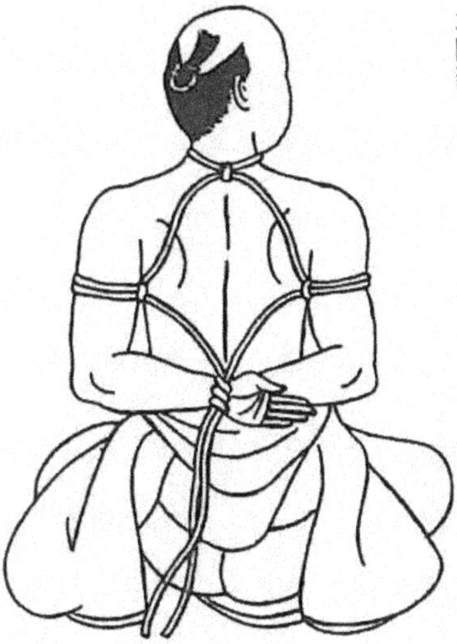

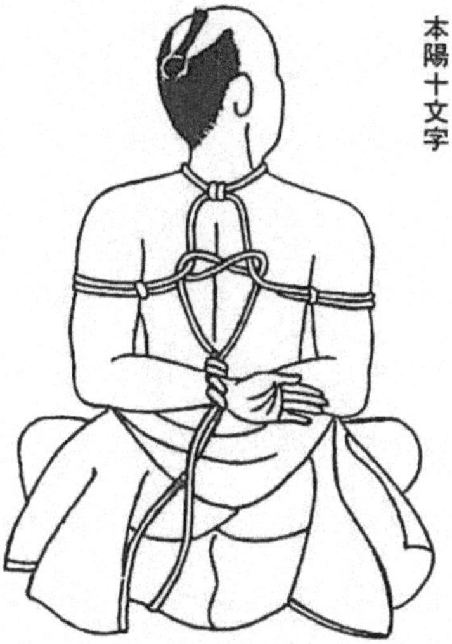

Ho-en School	
Sen-O Kata Shikomi: In the Shape of our Long Past Lord	Haya Saru Musubi: Fast Monkey Tie
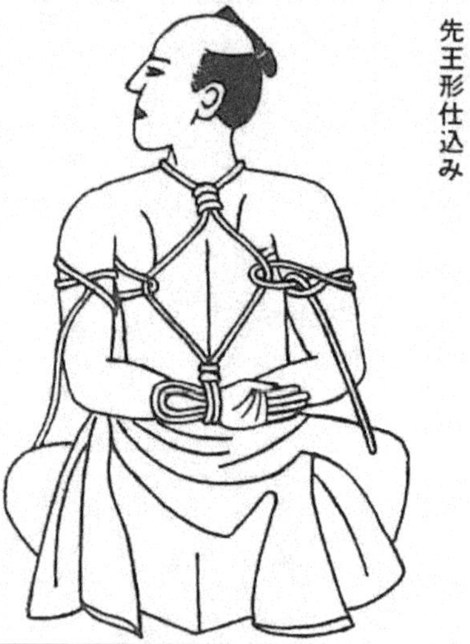 先王形仕込み	 早猿結び
Hiki-Watashi Kusari Gake: Handing Over Chain Tie	Haya Kumo Ito: Fast Spider's Web
 引渡鎖掛け	 早蜘蛛絲

Ho-en School

Nagasode Uruo Gata:
Long Sleeved Fish Scale Shape

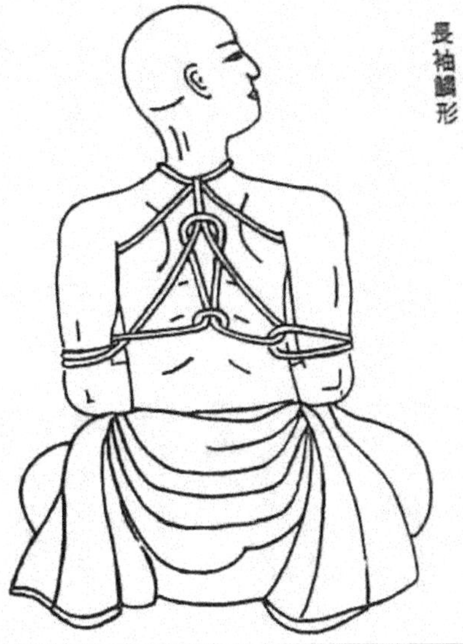

Onna Goho:
Five Different Women

Seigo Ryu

制剛流

Kajiwara Ryu

梶原流

Hojo Jutsu

捕縄術

Bondage Techniques of the Seigo School

Kajiwara School

Rope Scroll of the Seigo School
The Matagen Book

制剛流縄之巻又玄集

五法　落花　千鳥
十文字　村雲　六道
籠破　船中　四海
羽替付　微塵

五法	Goho	Five Rules
落花	Rakka	Falling Flower
千鳥	Chidori	Plover (a bird)
十文字	Jumonji	Cross Shaped
村雲	Murakumo	Cloudy Village
六道	Rokudo	Six Paths
籠破	Kago Yaburi	Breaking the Basket
船中	Funachu	On a Boat
四海	Shikai	Four Seas
羽替付	Hagai Tsuke	Donning a Cloak of Shed Feathers
微塵	Mijin	Pulverize

Seigo School
Kamosake Knot Illustration
Blue, Yellow, Red, White and Black Illustration

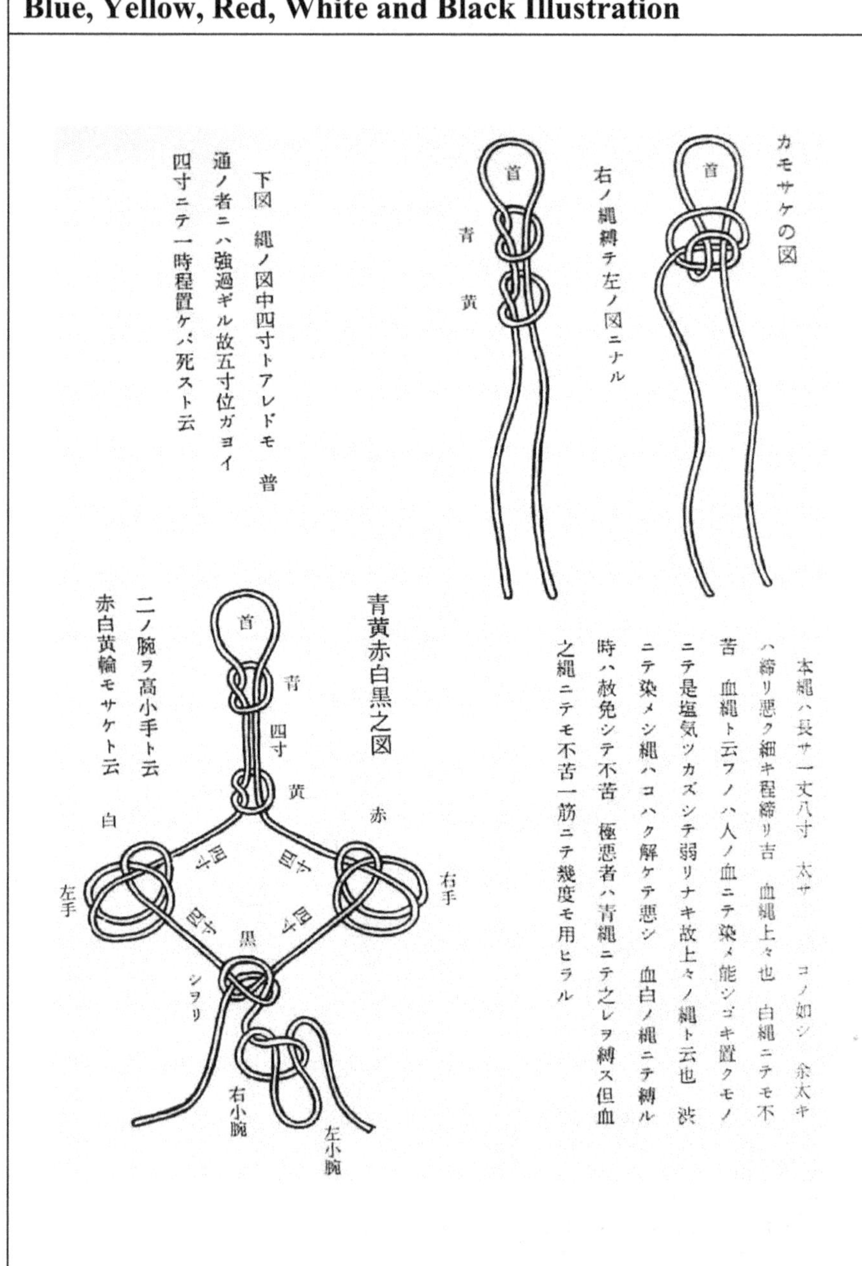

Seigo School
How to Tie the Kamosake Knot

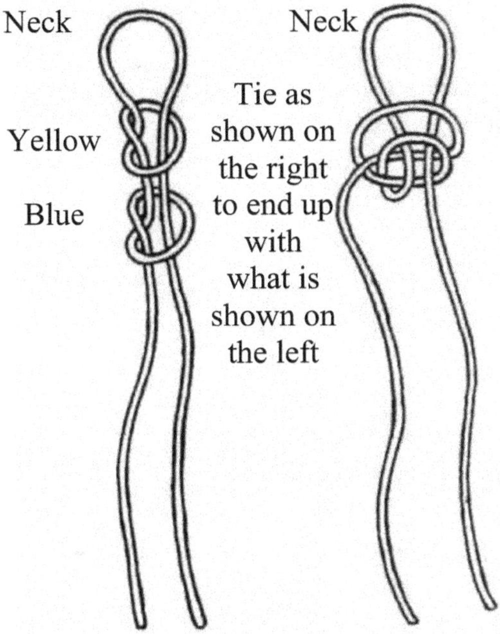

The rope used for Hon Nawa, or Main Tie, should be 1 Jo 8 Sun, 200 centimeters, in length with a thickness of [section missing.] If the rope is too thick it will be difficult to tie. A thinner rope makes for better ties. Blood Rope is the best. White rope causes the least suffering. Blood Rope is a rope that has been dyed with human blood. Rope dyed with human blood will not degrade over time due to the salt, therefore it has been considered the best rope since days of yore.

Blood Rope
Rope dyed with astringents are rough and are difficult to untie. Blood-White Rope is tied when a prisoner is being released from jail, so it is not painful. Truly evil prisoners are tied with blue rope. It is important to note that even if you tie multiple loops Blood Rope is not painful.

Seigo School
Blue, Yellow, Red, White and Black Illustration

The illustration below shows an interval between knots of 4 Sun (pronounced “soon”) 12 centimeters. This is too tight for most people. Typically when tying this restraint an interval of 5 Sun, 15 cm is used. Leaving a person tied with 12 cm sections for a long time will result in their death.
Another word for Ni no Ude, or bicep, is Taka Kote, Upper Forearm. The Red, White and Yellow loops are also called Sake.

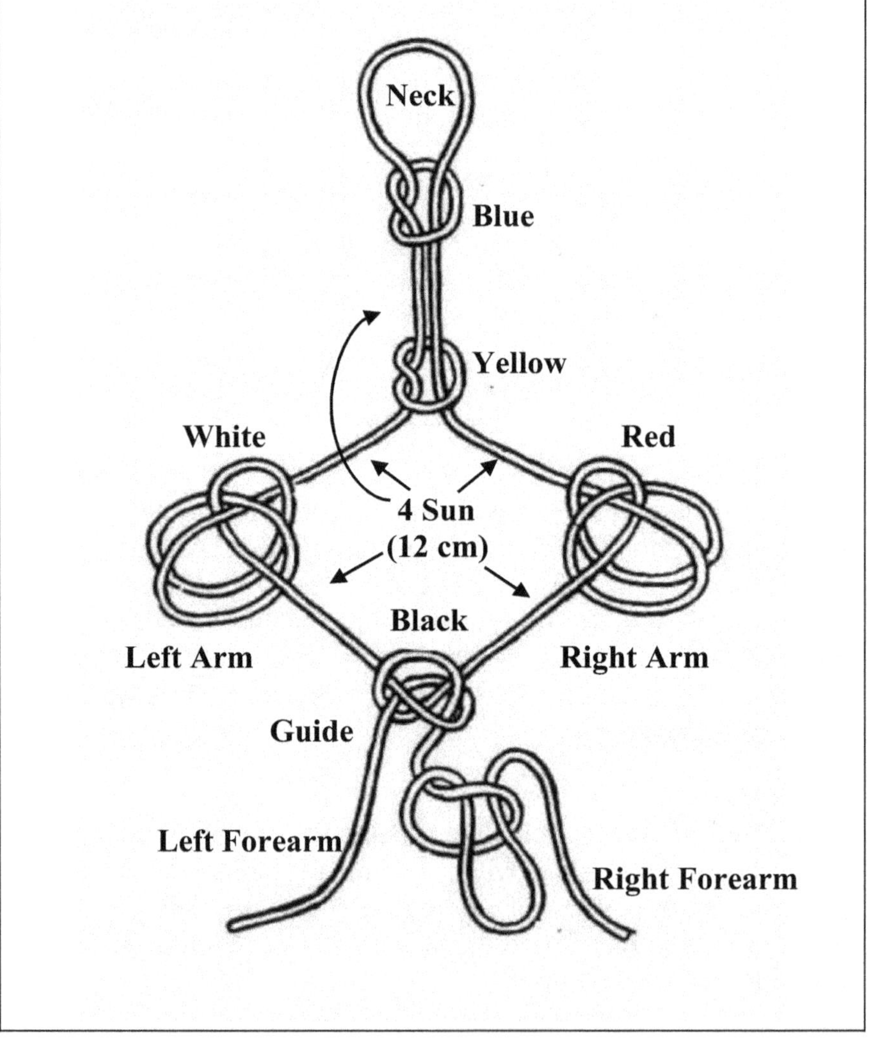

Seigo School	
Goho: Five Ways #2 **For use on:** **Priests** **Mountain Ascetics** **Women**	**Goho: Five Ways #1** **For use on:** **Peasants** **Townsfolk** **Mid-Level Officials**

五法

百姓
町人
中間

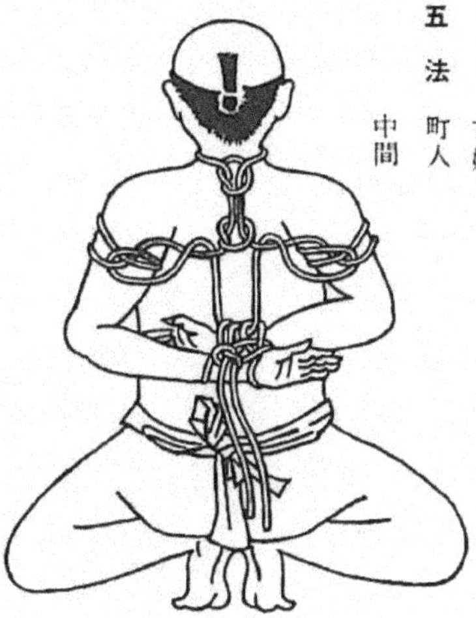

上図五法ハ赤白ノ縄余リヲ 黄
ト赤トノ間の縄ノ下ヘ図ノ如ク取
テ黒ヲ結ビ二重に廻シ割縄ヲ懸
上ヲシヲリニ結ブ
下図ハ二重回シ割縄裏ノ図

下図五法ハ赤白ノ縄ノ余リヲ直
ニ取 黒ノ結ビ一重宛廻シ割縄ナ
クシヲリニ結ビ其余リヲ腹ヘ廻シ
テ置ケバ 手上ヘ揚ラズ 長袖ノ
類 比縄ガヨイ
下図ハ割無縄裏ノ図

五法

坊主
山伏
女

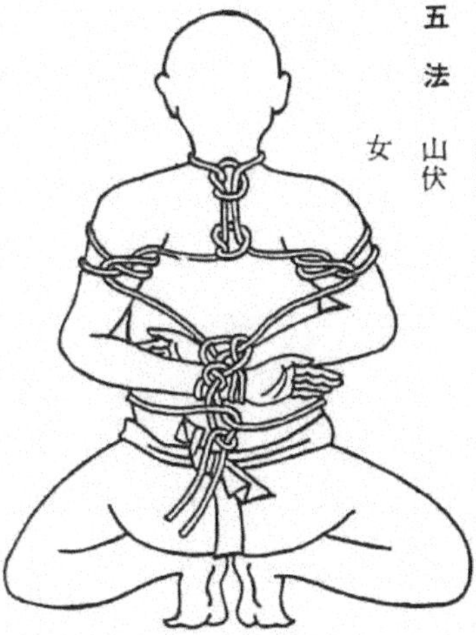

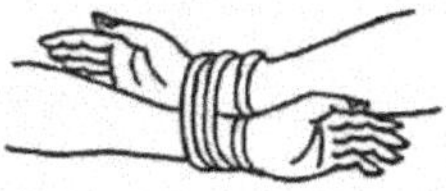

Seigo School
Goho: Five Ways #1
For use on: Peasants, Townsfolk & Mid-Level Officials

This illustration of Goho shows the remaining rope from the White and Red sections being used to tie the hands. The hands are tied between Yellow and Black*. The hands are tied as shown in the bottom illustration. A knot is tied at Black and the rope is wrapped between the hands twice in a Wari-Nawa, with one strand wrapping clockwise and the other wrapping counter-clockwise. A finishing knot is tied last.

*Note: The text says "between yellow and red" but this is probably a transcription error.

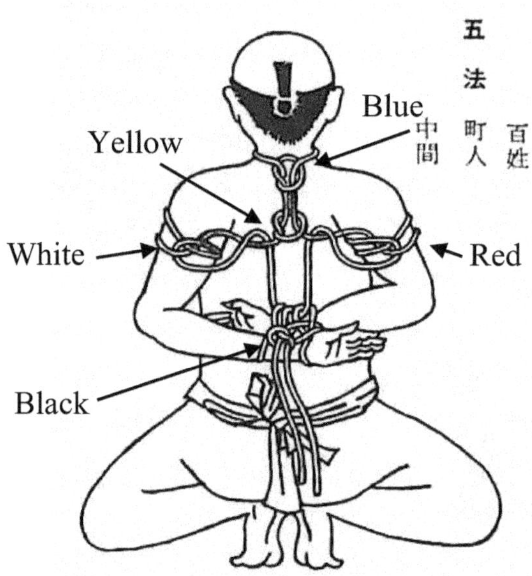

The bottom illustration shows the hands from the other side. The hands have been tied and the Wari Nawa tie is being wrapped twice between the hands, with each strand wrapping in a different direction.

Seigo School
Goho: Five Ways #2
For use on: Priests, Mountain Ascetics & Women

The illustration below shows how the rope remaining from Red and White is immediately used to tie Black. The rope is wound around one time and a Wari Nashi Nawa, No-Split Knot, is tied. Connect this to the remaining rope and wrap it around the stomach. If you do this the prisoner will have difficulty standing. This tie works well with long-sleeved clothing.

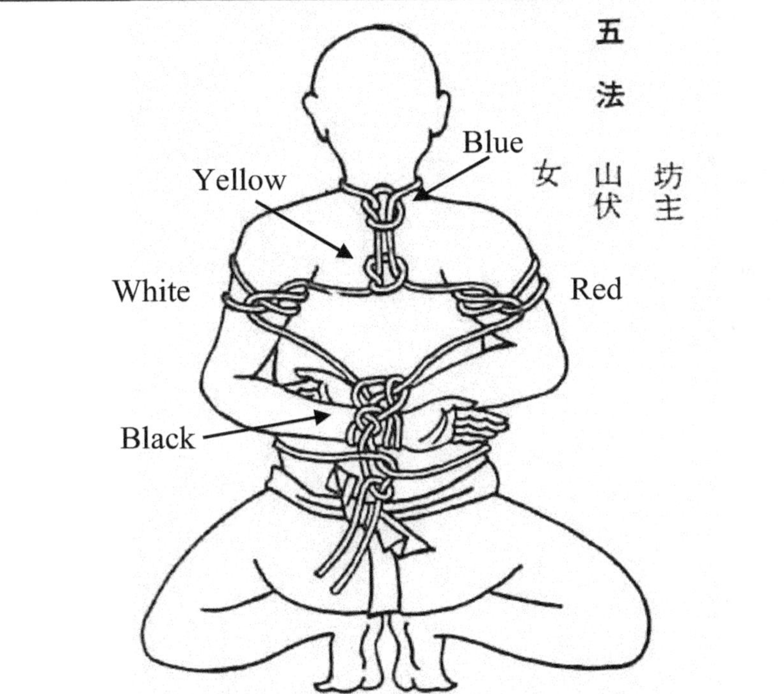

The illustration below shows what Wari Nashi Nawa, No-Split Rope, looks like when not covered by the tie at black.

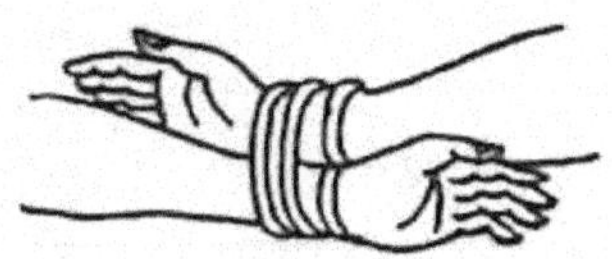

Seigo School

Rakka Kiri Nawa: Falling Flower Cut Tie (Executioner's Tie)
Rakka: Fallen Flower

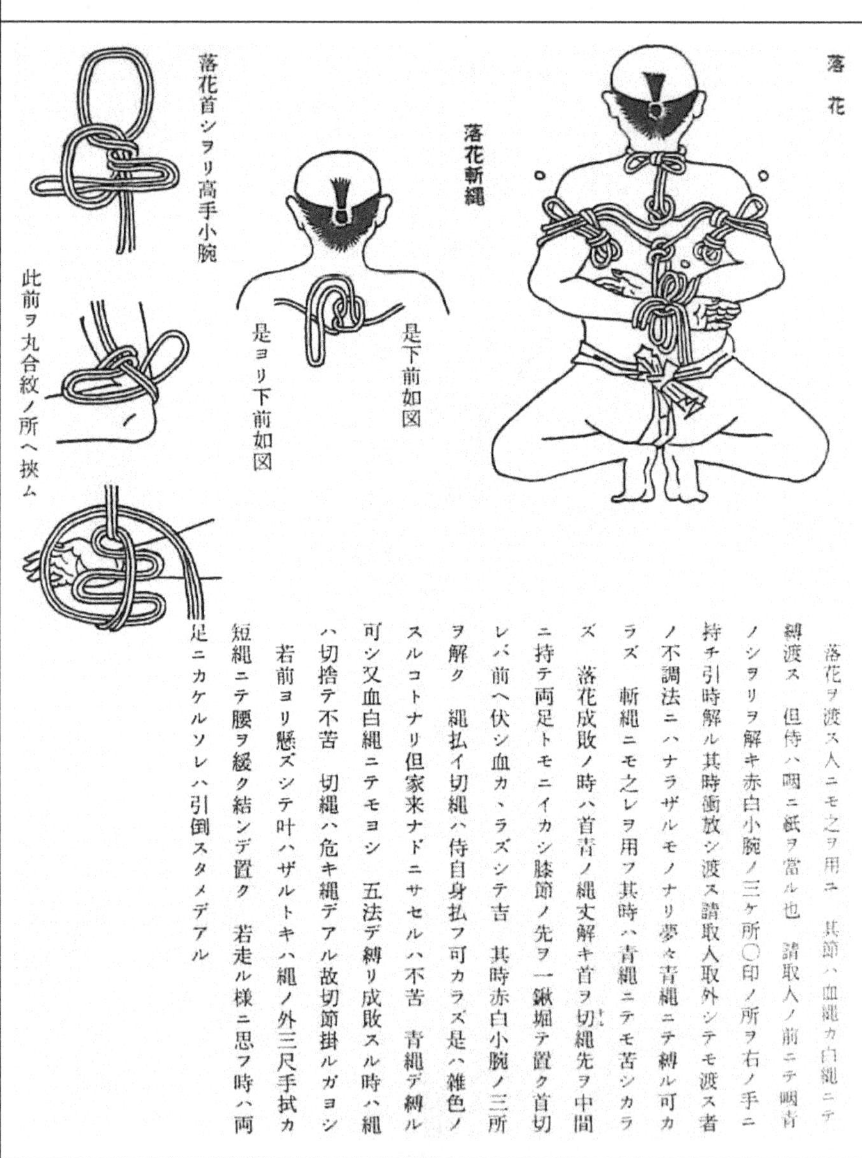

落花ヲ渡ス人ニモ之ヲ用ユ　其節ハ血縄カ白縄ニテ縛渡ス　但侍ハ咽ニ紙ヲ當ル也　請取人ノ前ニテ咽青ノシヲリヲ解キ赤白小腕ノ三ケ所〇印ノ所ヲ右ノ手ニ持チ引時解ル其時衝放シ渡ス請取人取外シテモ渡ス者ノ不調法ニハナラザルモノナリ夢々青縄ニテ縛ル可カラズ　斬縄ニモ之レヲ用フ其時ハ青縄ニテモ苦シカラズ　落花成敗ノ時ハ首青ノ縄丈解キ首ヲ切縄先ヲ中間ニ持テ両足トモニイカシ膝節ノ先ヲ一鍬堀テ置ク首切レバ前ヘ伏シ血カヽラズシテ吉　其時赤白小腕ノ三所ヲ解ク　縄払イ切縄ハ侍自身払フ可カラズ是ハ雑色ノスルコトナリ但家来ナドニサセルハ不苦　青縄デ縛ル可シ又血白縄ニテモヨシ　五法デ縛リ成敗スル時ハ縄ハ切捨テ不苦　切縄ハ危キ縄デアル故切節掛ルガヨシ　若前ヨリ懸ズシテ叶ハザルトキハ縄ノ外三尺手拭カ短縄ニテ腰ヲ緩ク結ンデ置ク　若走ル様ニ思フ時ハ両足ニカケルソレハ引倒スタメデアル

Seigo School

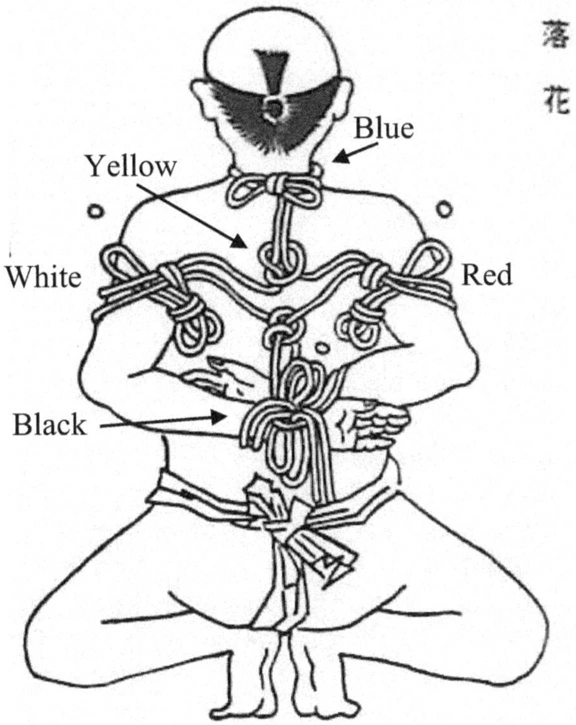

落花ヲ渡ス人ニモ之ヲ用ユ　其節ハ血縄カ白縄ニテ縛渡ス　但侍ハ咽ニ紙ヲ當ル也　請取人ノ前ニテ咽青ノシヲリヲ解キ赤白小腕ノ三ケ所〇印ノ所ヲ右ノ手ニ持チ引時解ル其時衝放シ渡ス請取人取外シテモ渡ス者ノ不調法ニハナラザルモノナリ夢々青縄ニテ縛ル可カラズ　斬縄ニモ之レヲ用フ其時ハ青縄ニテモ苦シカラズ　落花成敗ノ時ハ首青ノ縄丈解キ首ヲ切縄先ヲ中間ニ持テ両足トモニイカシ膝節ノ先ヲ一鍬堀テ置ク首切レバ前ヘ伏シ血カ、ラズシテ吉　其時赤白小腕ノ三所ヲ解ク　縄払イ切縄ハ侍自身払フ可カラズ是ハ雑色ノスルコトナリ但家来ナドニサセルハ不苦　青縄デ縛ル可シ又血白縄ニテモヨシ　五法デ縛リ成敗スル時ハ縄ハ切捨テ不苦　切縄ハ危キ縄デアル故切節掛ルガヨシ

若前ヨリ懸ズシテ叶ハザルトキハ縄ノ外三尺手拭カ短縄ニテ腰ヲ緩ク結ンデ置ク　若走ル様ニ思フ時ハ両足ニカケルソレハ引倒スタメデアル

Seigo School
Rakka: Fallen Flower / Executioner's Tie /Banishment Tie

Rakka is a tie used when transferring prisoners. Typically either Blood rope or White rope is used to secure them for transfer. However, if the prisoner is a Samurai then a piece of paper should be wrapped around the neck. When standing before the person taking custody of the prisoner, first untie Blue. Next, take hold of the places marked with a ○, the Red, White and Black ties, with your right hand. At the moment of transfer, pull all three places together, releasing the ties. Allowing the person receiving custody of the prisoner to undo the ties is improper. Ensure you do not leave the blue rope tied.

When using the Fallen Flower tie as a Kiri Nawa, Executioner's Knot, ensure the knot at Blue is not excessively tight. Success or failure with Fallen Flower is determined is when the Blue rope around the neck is untied and the prisoner is beheaded. You should hold the remaining rope in the middle and keep both legs braced.

Before the execution, scrape a small trench in the dirt with a hoe in front of the prisoner's kneecaps. This will cause him to fall forward after being cut and you won't be covered in blood. After the execution remove the ropes from the 3 points: Red, White and Black.

Executioner's Knot can be used for Nawa Harai, a person sentenced to being banished while tied, however remember this shouldn't be used for Samurai. General purpose rope should be used. The exception to the rule is for Samurai retainers. The color of the rope should be blue, but white is also acceptable.

Success or failure when tying Goho is determined if you run out of rope or not. It is easy to run out of rope.

Executioner's Knot can be dangerous; therefore you should be sure to tie the rope firmly. If, as mentioned before, you are unable to complete the tie with the amount of rope you have, use a 3 Shaku (90cm) Tenugui (all-purpose cloth) or a short piece of rope to tie the prisoner to your waist.

If you think the prisoner may try to run, you can also tie this around their legs to pull them down when they try to escape.

Seigo School	
Rakka Kiri Nawa: Falling Flower Executioner's Tie **Rakka: Fallen Flower / Executioner's Tie**	
1. The sequence continues in illustrations 2 ~ 4.	2. Diagram of how the Taka Kote, or biceps, should be tied off after the neck is tied in Rakka.
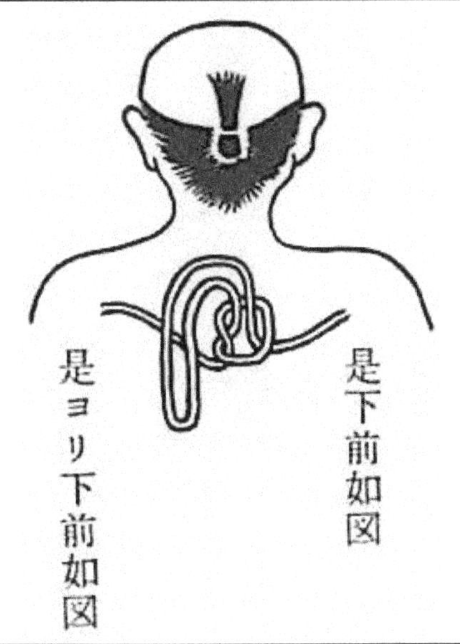 	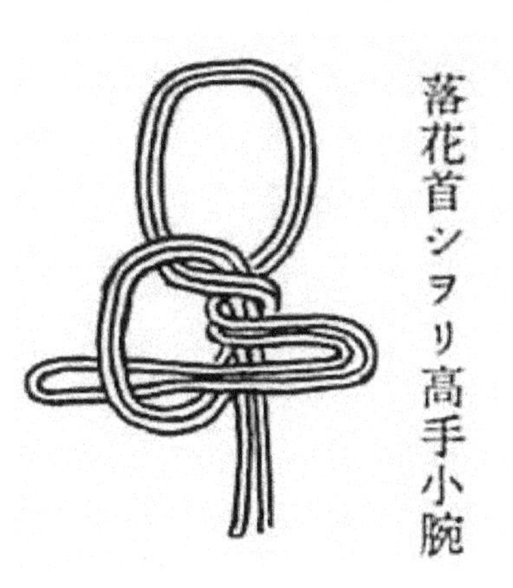
3.The loop is secured by wedging it under itself like this.	4. [no text]
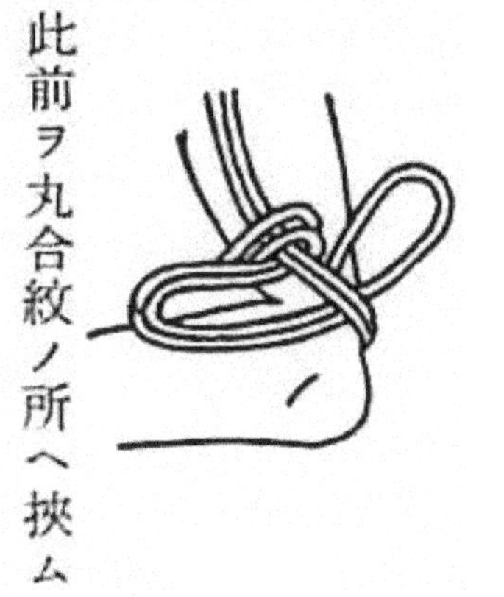 	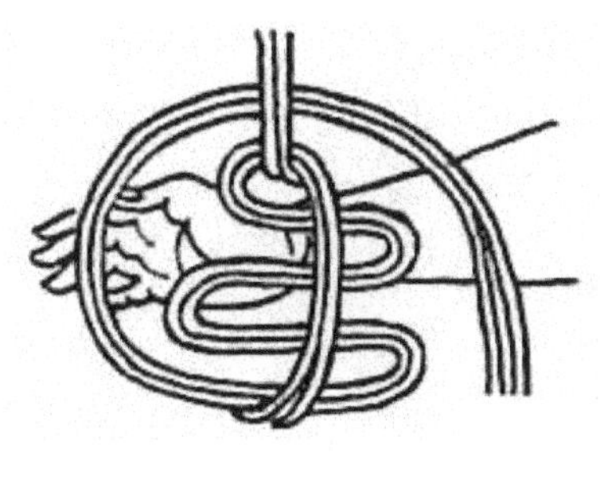

Seigo School	
Jumonji Daihiji: Great Secret Cross Shaped Tie	**Chidori: Plover Tie**

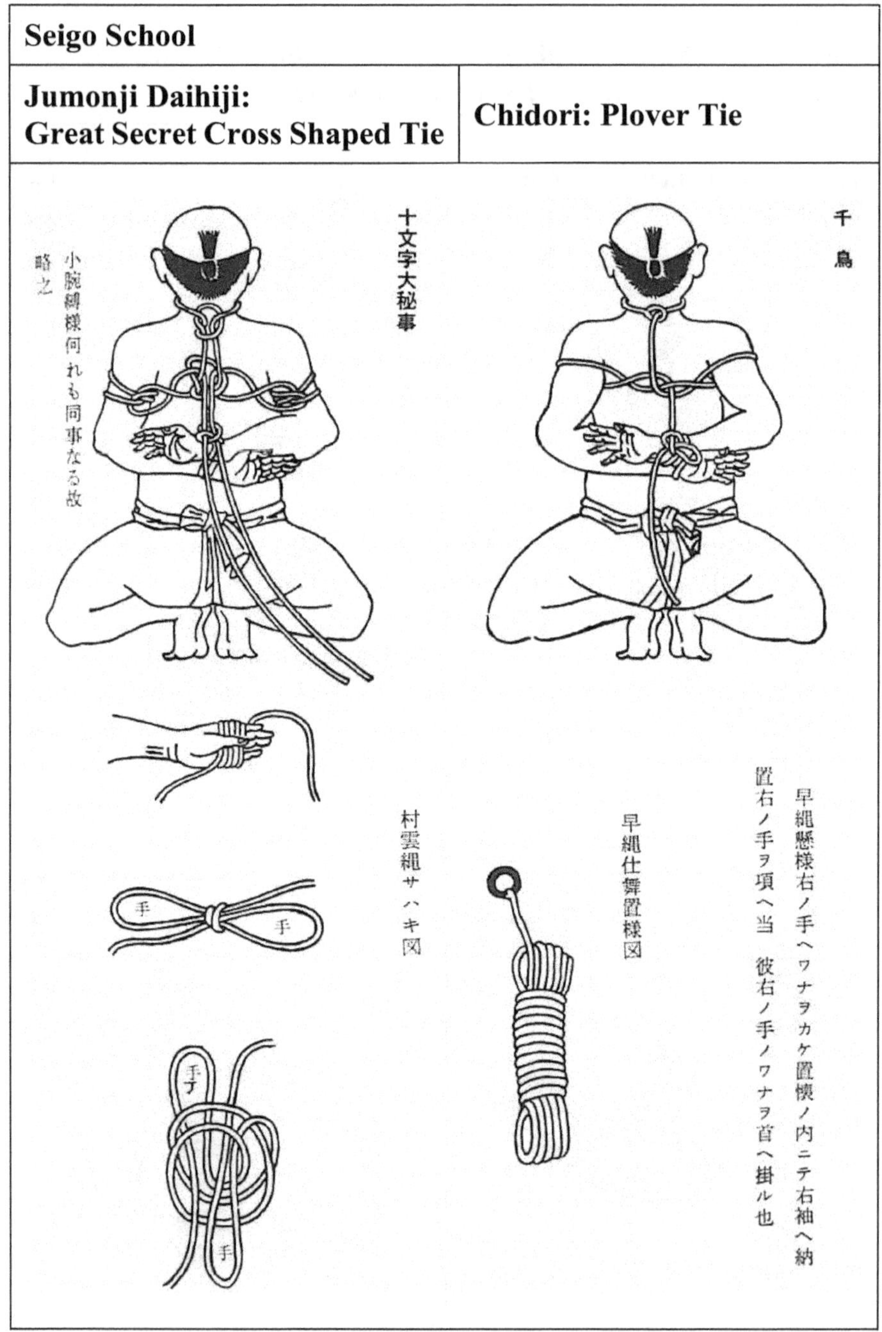

Seigo School	
For Haya Nawa, Fast Tie, first loop the ring around your wrist and stuff the rope up your sleeve. Place your right hand at the base of the prisoner's neck and wrap the cord around his neck.	
Chidori: Plover Tie (No description)	Haya Nawa

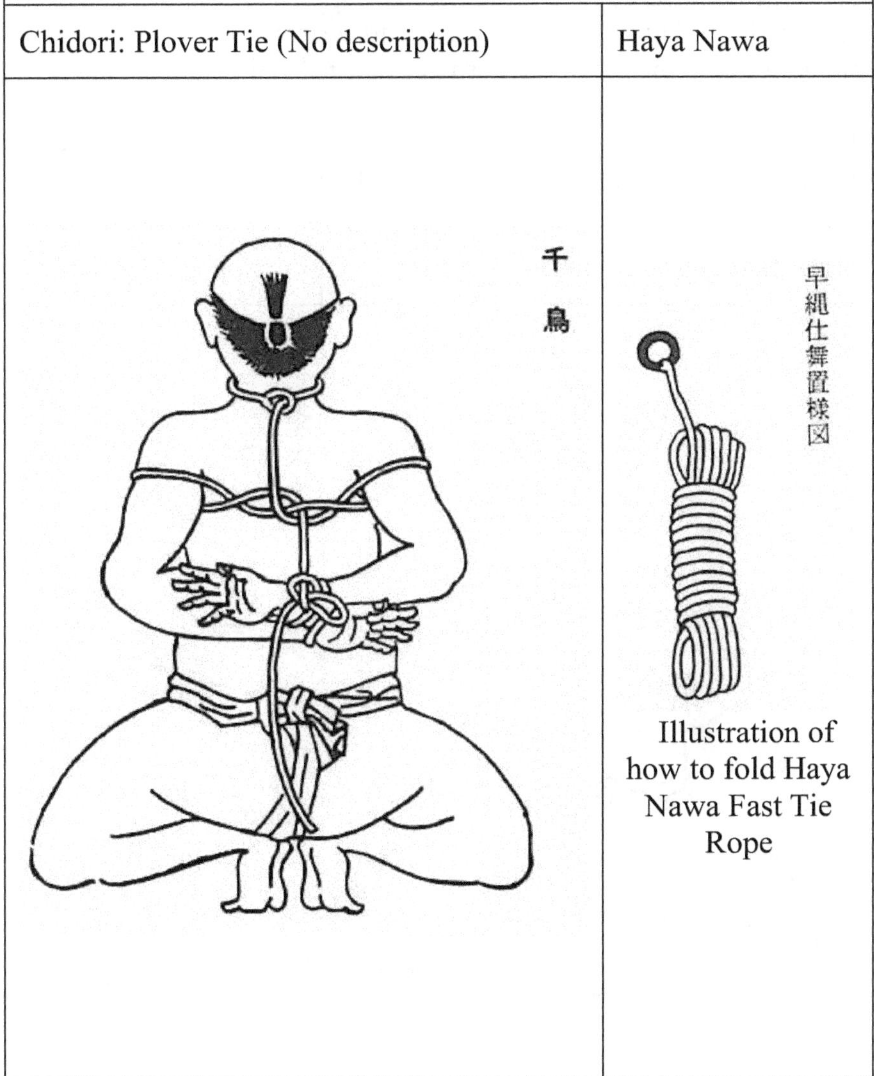

Illustration of how to fold Haya Nawa Fast Tie Rope

<table>
<tr><td colspan="2">Seigo School</td></tr>
<tr><td colspan="2">Left Illustration:
Jumonji Daihiji: Great Secret Cross Shaped Tie
Illustration of how the upper arm and wrists should be tied. As the method is the same further instructions will be abbreviated.

Right Illustration:
How to tie the Mura Kumo Cloudy Village Knot</td></tr>
<tr><td>Great Secret Cross Shaped Tie</td><td>Cloudy Village Knot</td></tr>
<tr><td colspan="2">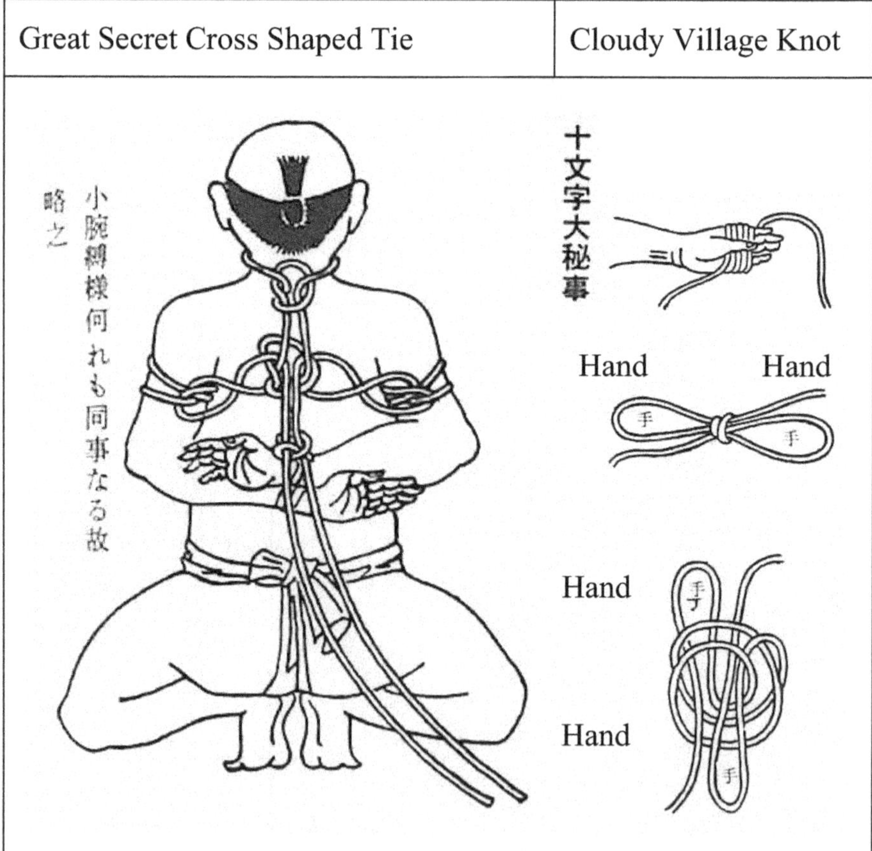
</td></tr>
</table>

Seigo School

Mura Kumo: Cloudy Village

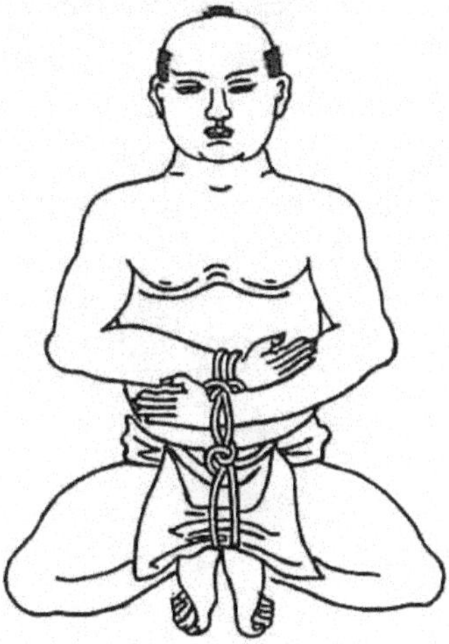

村雲

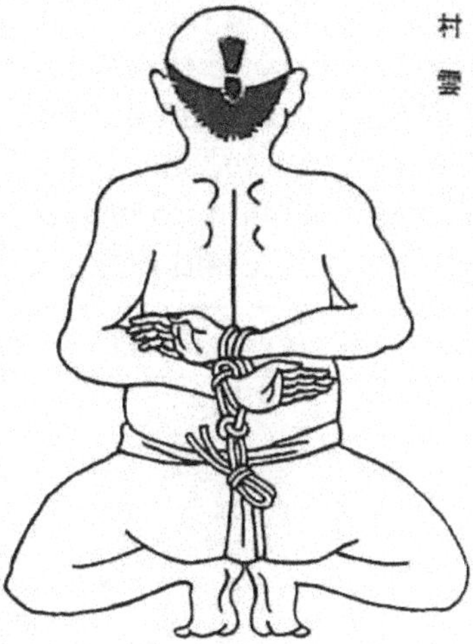

村雲

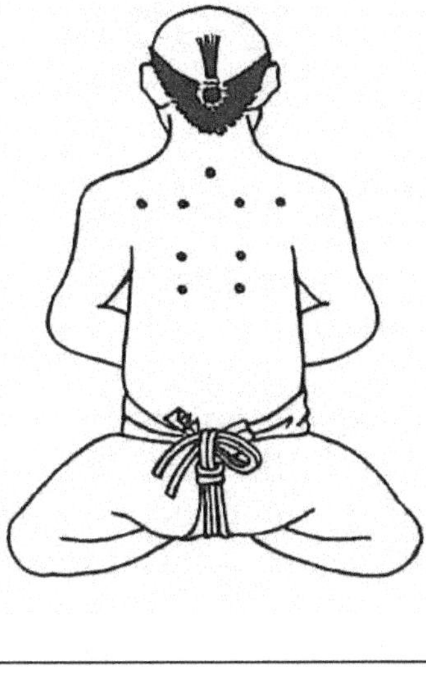

村雲

上図　村雲　後ニテ縛ル時ハ前ヘ廻スニ及バズ　褌ヲ能シメ三ツカラミニ能結ビ付テ置ク　褌ヲシメル儀口伝ナリ

次村雲　前ニテ縛ル時ハ股ヨリ後ノ方ヘ廻シ褌ニ結ビ付ル　後ヘ廻サズニオケバ口ニテトクカラデアル

Seigo School

Mura Kumo: Cloudy Village
The illustrations show Mura Kumo. When tying the hands behind the back, make sure the prisoner's Fundoshi (loincloth) is secured firmly and that the knot can't be shifted to the front. The three tie points should be done securely. There is a Kuden, oral tradition, regarding how to firmly tie the Fundoshi. The illustration on the top left shows Mura Kumo tied in the front. The strands of rope should be passed under the thighs and tied at the back of the Fundoshi. If you do not tie the rope behind the prisoner, he will feel free to talk and plead his case.

Cloudy Village tied on thc front	Cloudy Village tied on the back
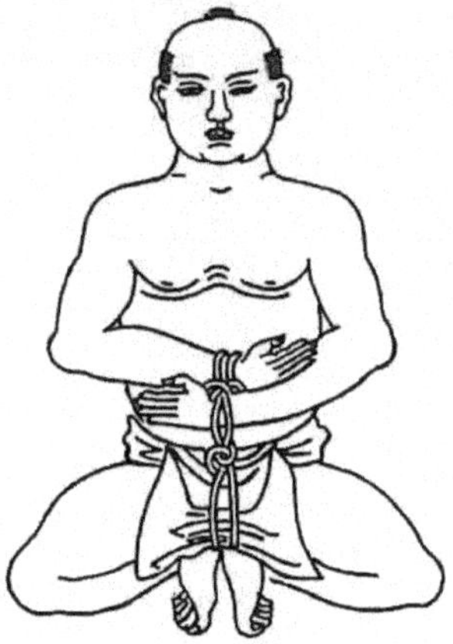	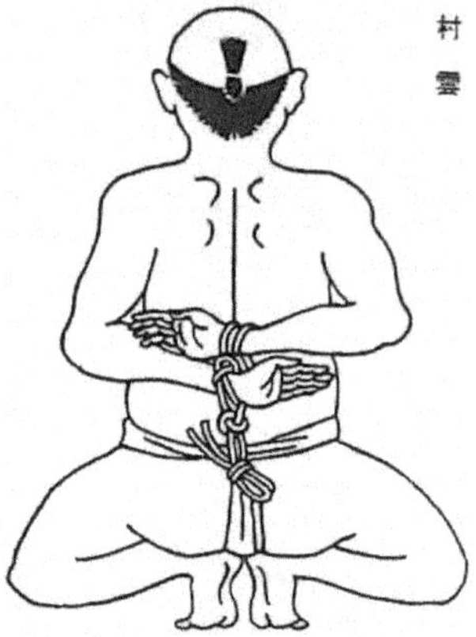 村雲
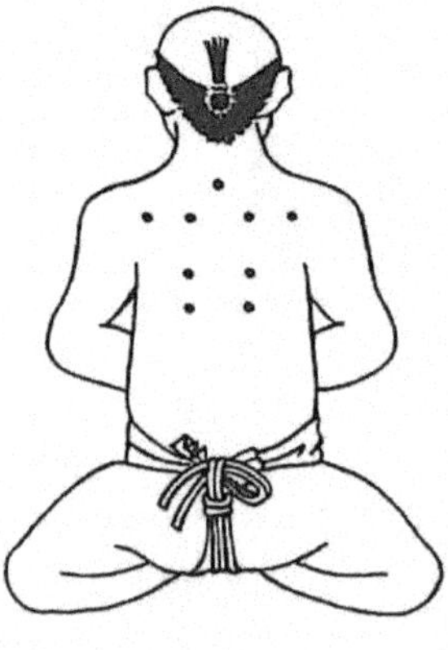	Left: Cloudy Village Translator's Note: This is showing how the knot should be tied in the back when the hands are tied in the front. There is no notation regarding the spots on the back. Possibly those are indicators showing where other ties should be done.

Seigo School
Rokudo-Shi: Six Paths of The Samurai
Senchu: Whilst on a Boat
Kago Yaburi: Breaking the Basket

六道士

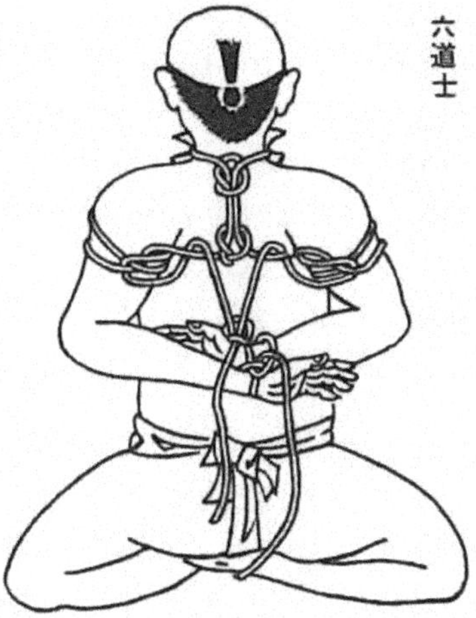

六道士ハ赤白ノ縄ノ余リヲ黄トノ間ノ縄下ヨリ上ヘ
取リ黒ノシヲリ仕咽計ニ紙一帖程当ル　是レ縄目ノ恥
辱ユヘ如此五法ノ内ト云ヘリ

籠破

籠破ハ小腕ヲ縛腹ヘ廻シ赤シヲリヲ結其縄ノ余リヲ
後ヨリ股ヘ通シ前ノ方ニテ帯ヘ通シ拟畳能真中ニ穴ヲ
明縄ヲ通シ下ニテ留ル或ハ後ヨリ直ニ下ヘ通シ留テモ
吉　何縄ニテモ如斯ナル故ニ図を略ス

船中

右船中何縄ニテモ腹ヘ廻シ船張ヘ結付置也依図ヲ略
ス

Seigo School

Rokudo-Shi Six Paths of the Samurai

Rokudo-shi, Six Paths of the Samurai Tie, uses the remaining rope from Red and White to pass through Yellow. The hands should be tied from the bottom up. Before tying Black make sure that one piece of paper is wrapped around the throat. This is because having a rope burn on the throat would be an insult. This is what is told in the story of Goho.

Note: The Six Realms of Karmic Rebirth **六道輪廻** you will be reborn. The Six Realms 六道 are:

天上 Heavenly – The highest rank
修羅 Demi-god – However in a state of warfare with others
人間 Human – Farmer, Craftsman, Merchant or Samurai
畜生 Animal – Reincarnated as a beast
餓鬼 Ghost - Food and drink torment but can't be reached
地獄 Hellish – A multitude of tortures

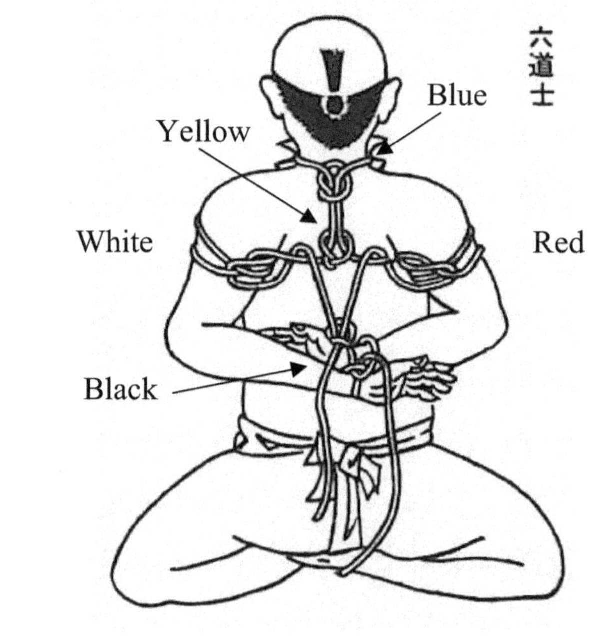

Seigo School

Senchu: Whilst on a Boat

When on a boat take any kind of rope, wrap it around the prisoner's waist and tie to a rail or the mast of the ship. This illustration will be abbreviated.

Kago Yaburi: Breaking the Basket

For Kago Yaburi, tie off the upper arms then wrap the rope around the body under the stomach. Take the rope remaining after tying Red and bring it under the crotch and pass through the front Obi (belt.) Fold this rope in half and make a loop. Finish by tying it off at the bottom or passing it back under the thighs again and tie it in the back. Since any tie can be used for this the detail illustration will be abbreviated.

Illustration from
An Illustrated Japanese Explanation of the Ten Kings of Hell
和解法華十王讃嘆絵抄

By Hatano Nichikyo 畑野日教 1883

Seigo School

Hagai Tsuke: Arm Lock
Shikai: Four Seas
Mijin: Pulverizer

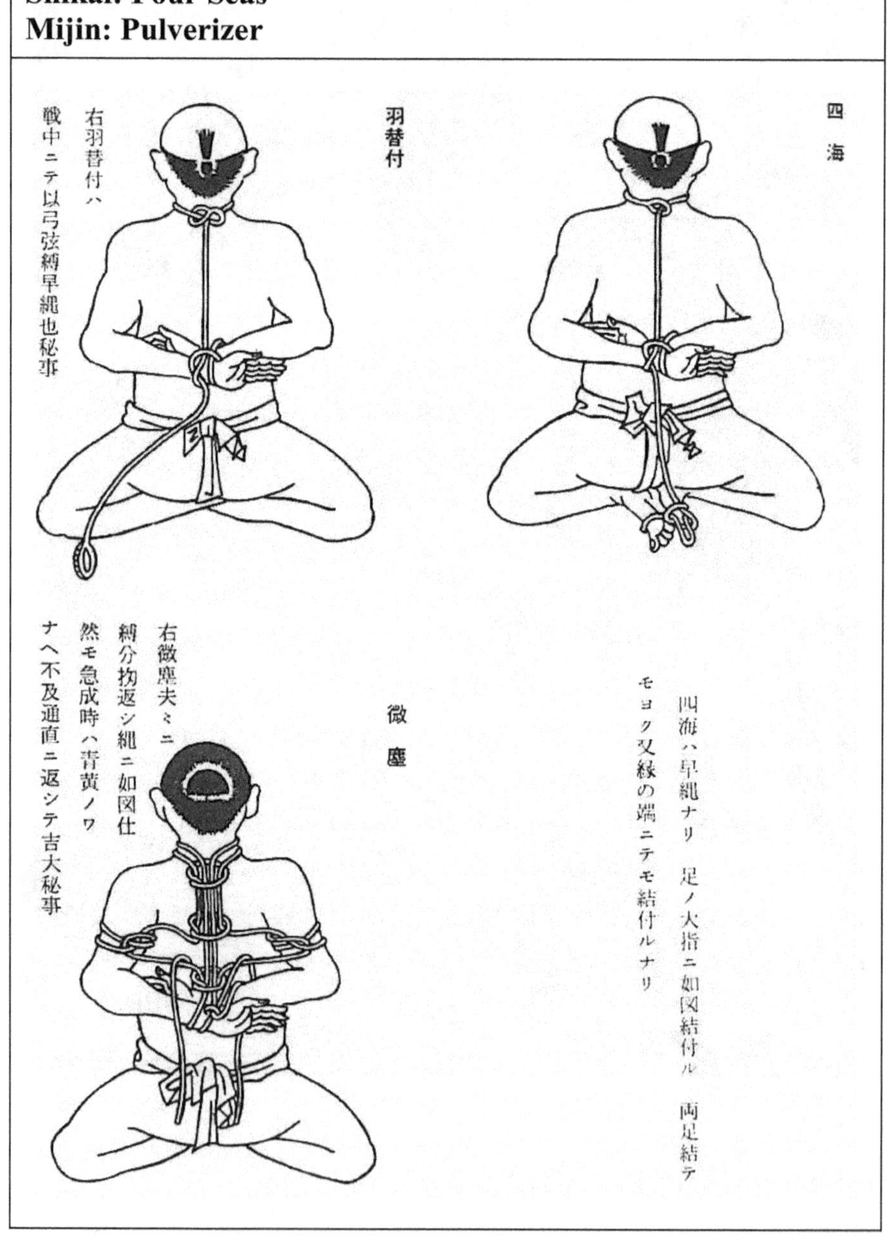

Seigo School

Shikai: The Four Seas (The Whole World)
Four Seas is a Haya Nawa, Fast Tie, technique. The big toe of the foot should be tied as shown in the illustration. Even though both legs are tied it is important to tie a knot at the end of the rope.

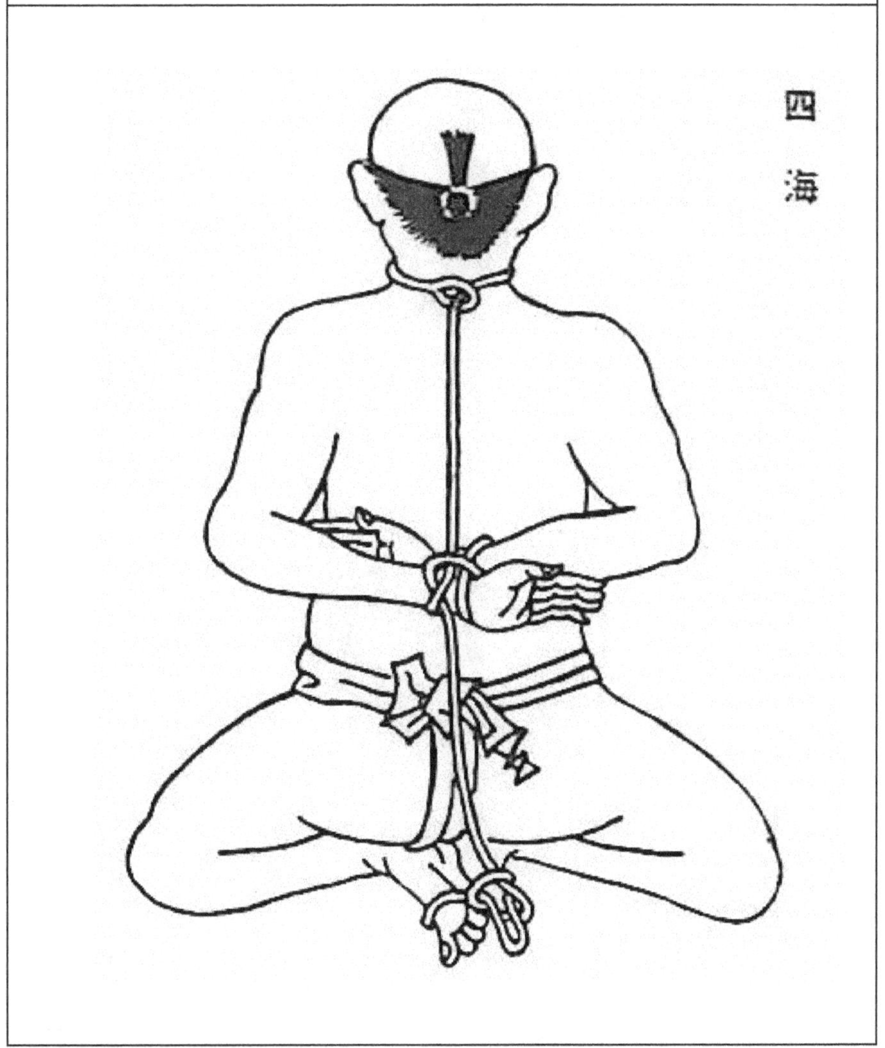

Seigo School
Hagai Tsuke : Reversed Wings Reversed Wings is used during battle. You have captured an enemy Samurai and used a bowstring to rapidly restrain him. This is a secret technique. Note: The loops around the end of the string are to hook onto the top and bottom of a bow.
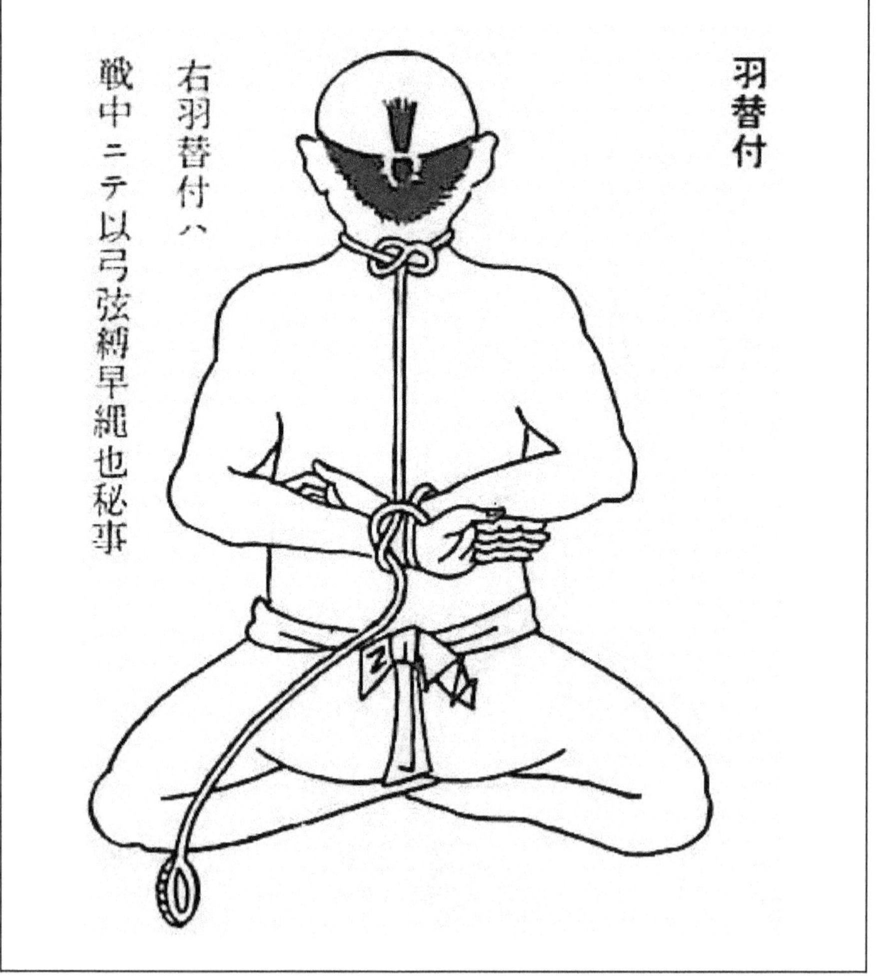

Seigo School

Mijin: Pulverizer

This illustration shows all the ties used in Mijin, Pulverizer. The points that the rope is doubled can be seen in the illustration. When tying in a hurry the rope proceeds straight down from blue and yellow with no loop for the upper arms. This is a great secret.

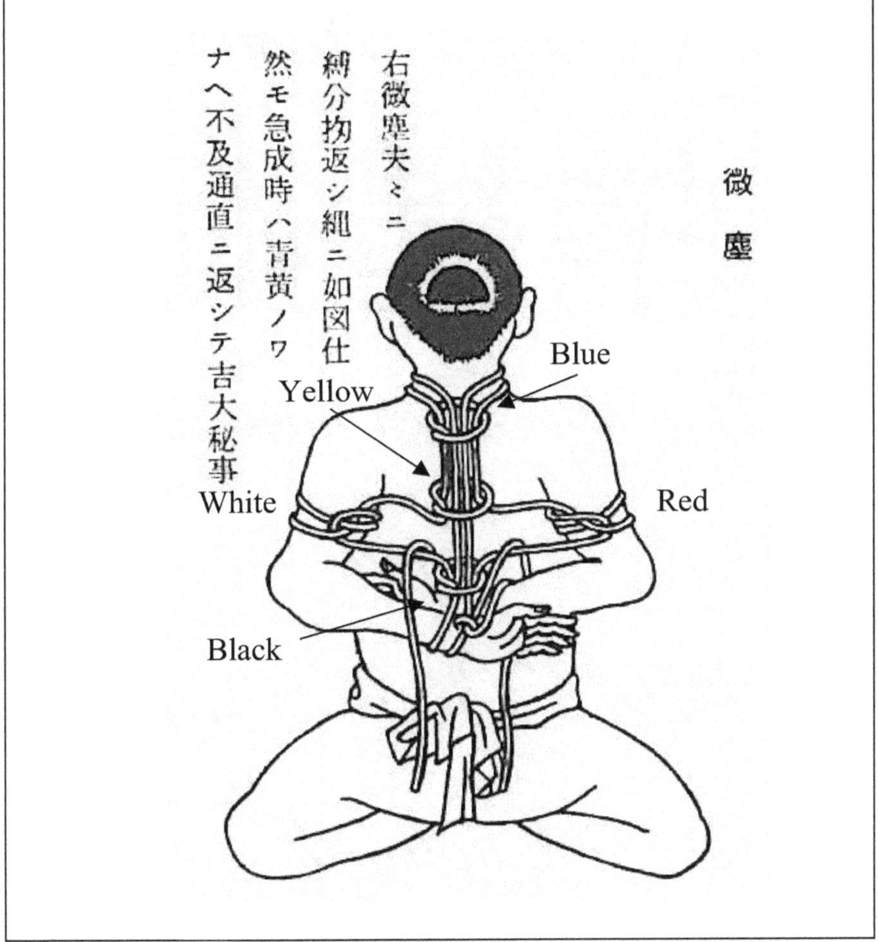

Igai Ryu

猪谷流

Hojo Jutsu

捕縄術

Bondage Techniques of the Igai "Wild Boar Valley" School

Igai "Wild Boar Valley" School Techniques

一　早縄　長さ六尺四寸　天二十八宿　地三十六儀を表せり　不動縛りの縄より初云掛様　蜘蛛の口伝

一　五法　常に掛る縄也　寸法七寸　但縄の長さ一丈二尺八寸　陰陽結口伝

一　千鳥　下﨟に掛る縄也　寸法七寸　高手に口伝　首陰陽結　七曜の星を表す口伝

一　村雲　児法師に懸也　寸法の伝　首根に陰陽結　高手に口伝　首根に紙巻口伝

一　十文字　諸囚人に懸る也　四方四寸海を表せり　但陰陽結東西南北口伝

一　船中　船中にて掛縄也　小手前後口伝

一　籠破　極意の縄也　陰陽結　高手小手に口伝　尤小手縄に返縄口伝

一　六道　侍に懸也、寸法六寸首に紙を巻へし陰陽結有高手習　是地獄　餓鬼　修羅　人天を表す
弓の弦にて懸る口伝

一　微塵　是は裘懸武者鎧の上より懸縄也　八曜九字十字十戒を表す縄也　弓弦二筋にて懸る也　第
一紫縄とは弓弦を云也　縄の付所に習有神前にては注連の縄　仏前にては前の網裘にては
胞衣を表せり陰陽結　東西南北口伝多し

一　落花　是は切縄也　首根引解にして高手に口伝　但　荒縄

Illustrated Guide to Igai School Bondage Techniques	
These Igai School techniques are derived from the Seigo School line as taught in the Kajiwara School. Note: The descriptions and illustrations were on different pages, so I have placed them on the same page	
五法	Goho: Five Methods
落花	Rakka: Fallen Flower
千鳥	Chidori: Plover
十文字	Jumonji: Cross Shaped Tie
村雲	Murakumo: Cloudy Village
六道	Rokudo: Six Paths
籠破	Kago Yaburi: Breaking the Basket
船中	Funachu: Whilst on a Boat
四海	Shikai: Four Seas
羽替付	Hagai Tsuke: Arm Lock
微塵	Mijin: Pulverize

Igai "Wild Boar Valley" School

Goho: Five Methods
Jumonji: Cross Shaped

猪谷流縄縛図

猪谷流は制剛流より出た梶原流より伝承の流派である

五法　十文字　千鳥
村雲　六道　落花
籠破　船中　微塵

Igai "Wild Boar Valley" School
Goho: Five Ways This is the most commonly used tie. The interval between each tie is 7 Sun (21cm.) The overall length of the rope is 1 Jo 2 Shaku 8 Sun (3.6 meters.) There is a Kuden concerning the Yin-Yang Knot. Note: 1 Jo = 3.03 meters

Igai "Wild Boar Valley" School
Jumonji: Cross Shaped This can be used for any type of prisoner. The interval between each tie should be 4 Sun (12 cm.) This represents the Shikai, Four Seas, that surround us. However for the Yin-Yang tie there is a Kuden regarding East, West, South and North. Note: The Kuden may relate to the proper direction to face, based on the season, when tying. Note: The illustrations are not in the same order as the Seigo School techniques, even though the names are identical, and the Igai School is related to the Seigo School.

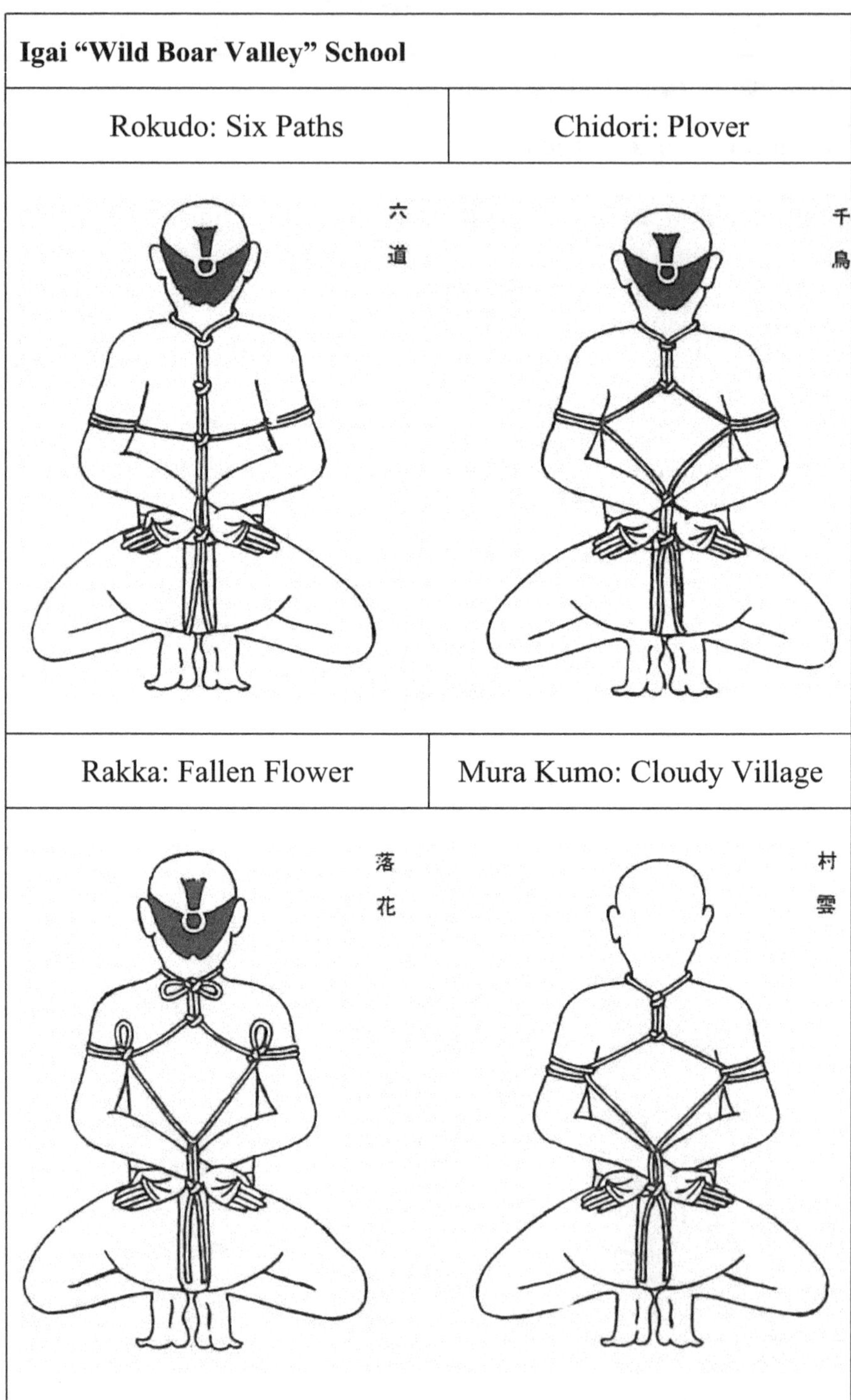
Igai "Wild Boar Valley" School
Rokudo: Six Paths
Chidori: Plover
六道
千鳥
Rakka: Fallen Flower
Mura Kumo: Cloudy Village
落花
村雲

Igai "Wild Boar Valley" School

Chidori: Plover

This is used on servants. The interval between each tie is 7 Sun (21 cm.) There is a Kuden about tying the Takate, Upper Arms. A Yin-Yang Knot should go around the neck. There is a Kuden about showing the Seven Luminaries (The 7 visible celestial bodies: Sun, Moon, Mercury, Venus, Mars, Jupiter and Saturn.)

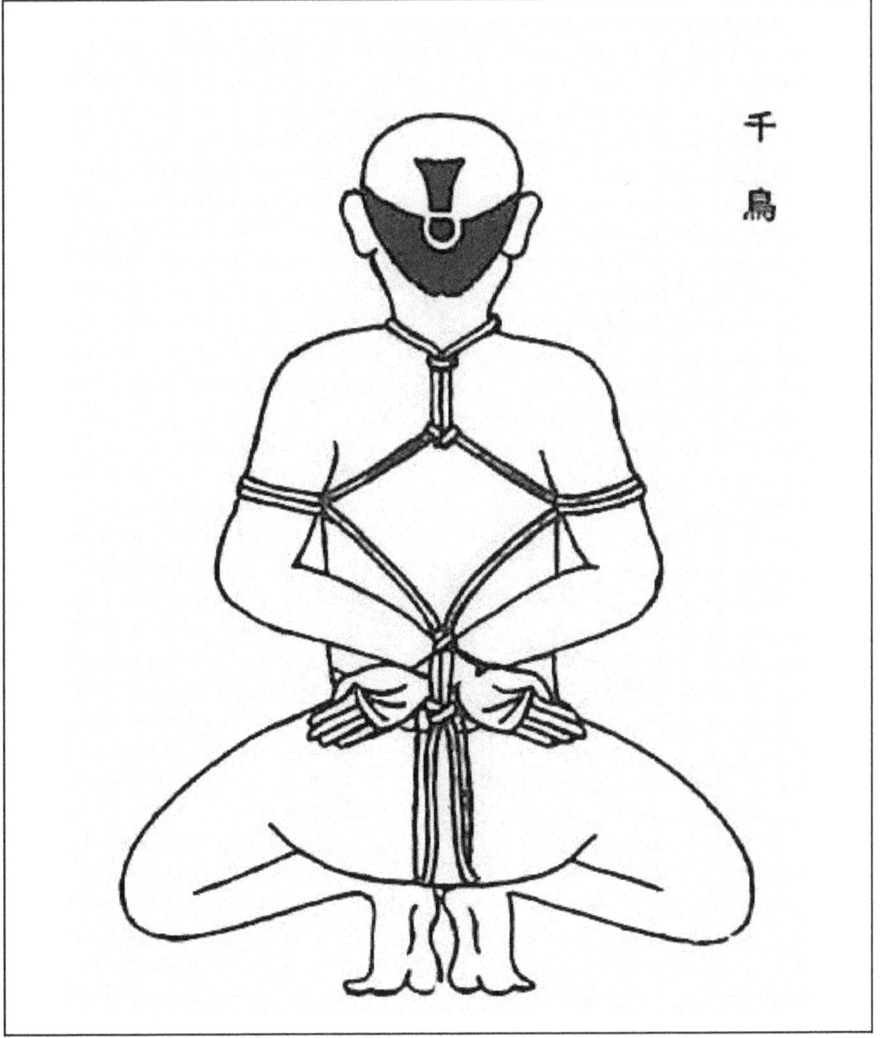

Igai "Wild Boar Valley" School
Mura Kumo: Cloudy Village This tie is used for children and Buddhist Monks. The length of the rope is a Kuden. A Yin-Yang Knot goes around the base of the neck. There is a Kuden about tying the Takate, Upper Arms. There is a Kuden about wrapping a piece of paper around the neck.
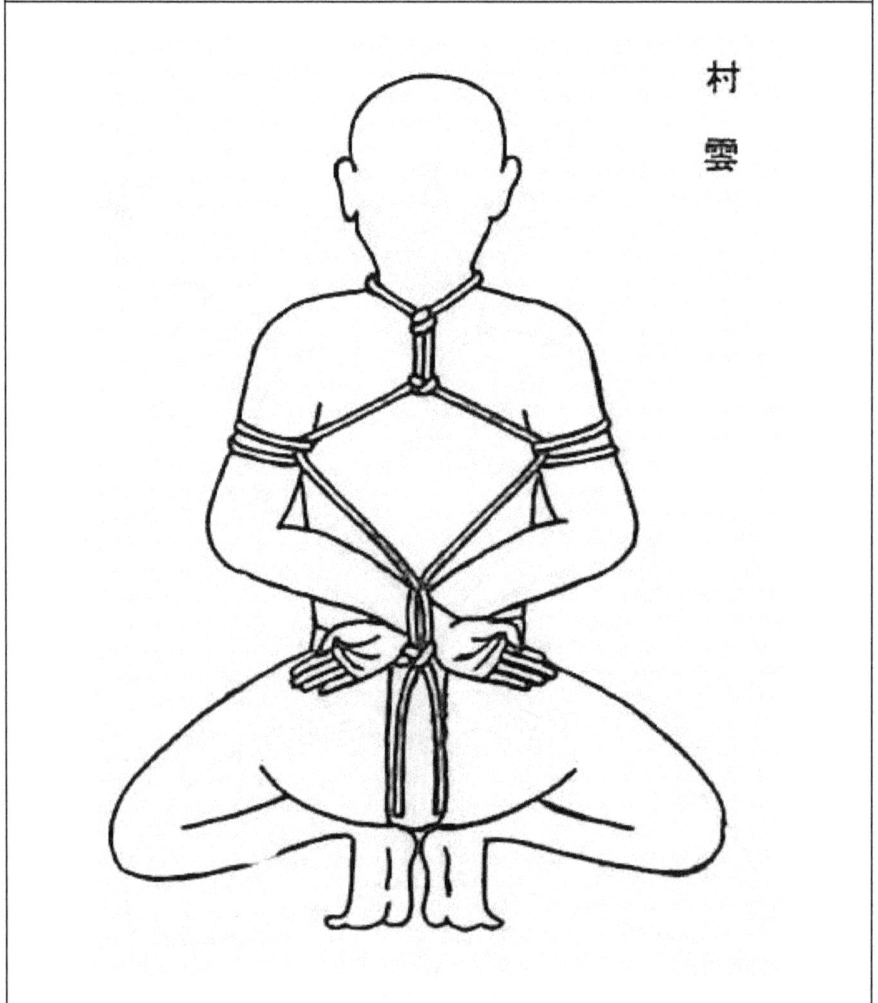

Igai "Wild Boar Valley" School
Rokudo: Six Paths This is used on Samurai. The interval between ties is 6 Sun (18cm.) You should wrap a piece of paper around the neck. The Yin-Yang Knot is used. There is a separate teaching about tying the upper arms. This tie represents: Hellish, (starving) Ghost, Demi-god and Heavenly.

Igai "Wild Boar Valley" School

Rakka: Fallen Flower

This is a Kiri Nawa, Cut Tie or Executioner's tie. The tie around the neck should be pulled free just before cutting. There is a Kuden about tying the upper arms, however it involves using rough or common rope.

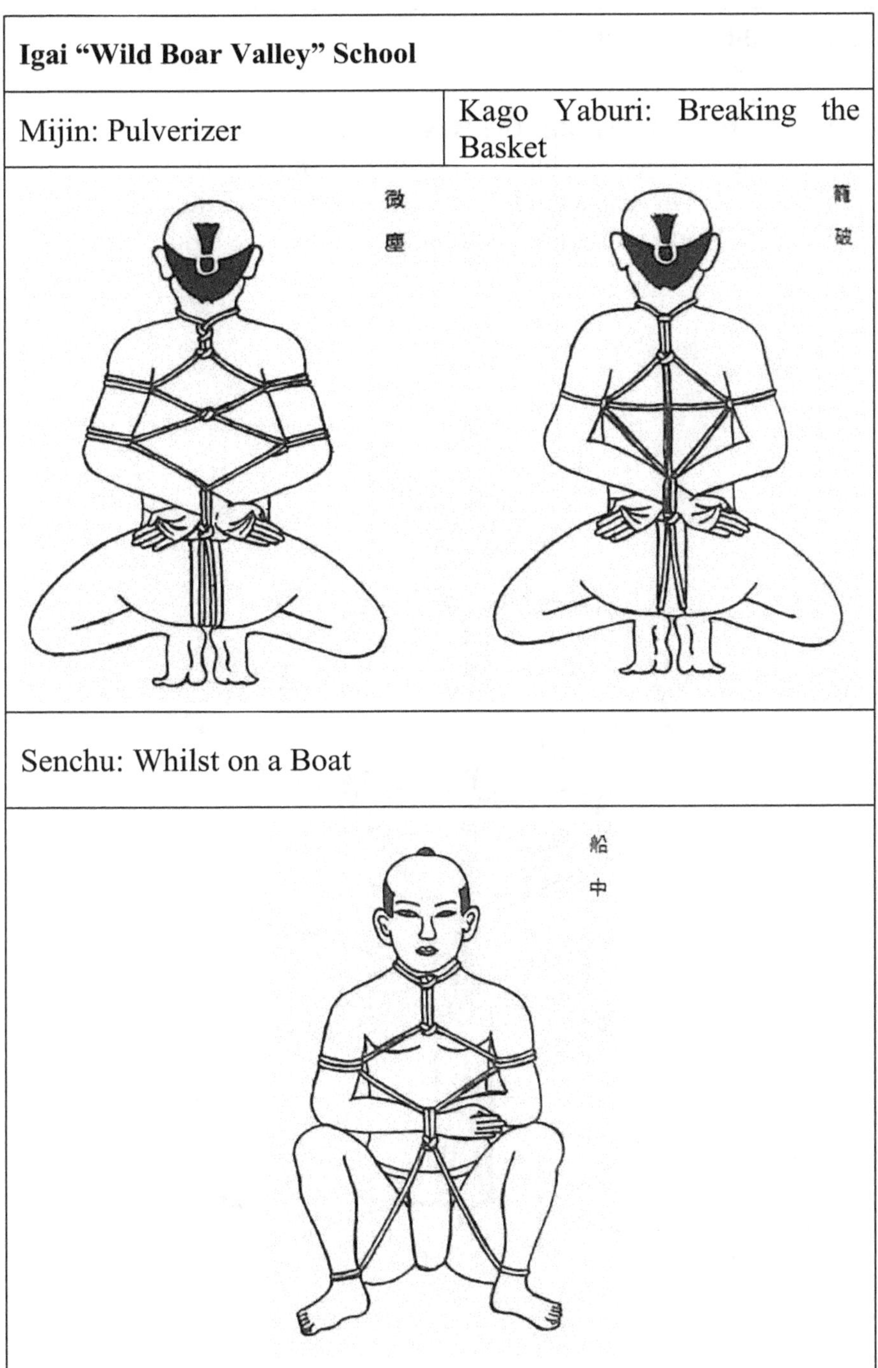

Igai "Wild Boar Valley" School	
Mijin: Pulverizer	Kago Yaburi: Breaking the Basket
Senchu: Whilst on a Boat	

Igai "Wild Boar Valley" School

Kago Yaburi: Breaking the Basket

This is the ultimate tie, the essence of this school. It uses the Yin-Yang Knot. There is a Kuden about the wrists and the upper arms. There is a Kuden about Kaeshi Nawa, Returning Knot.

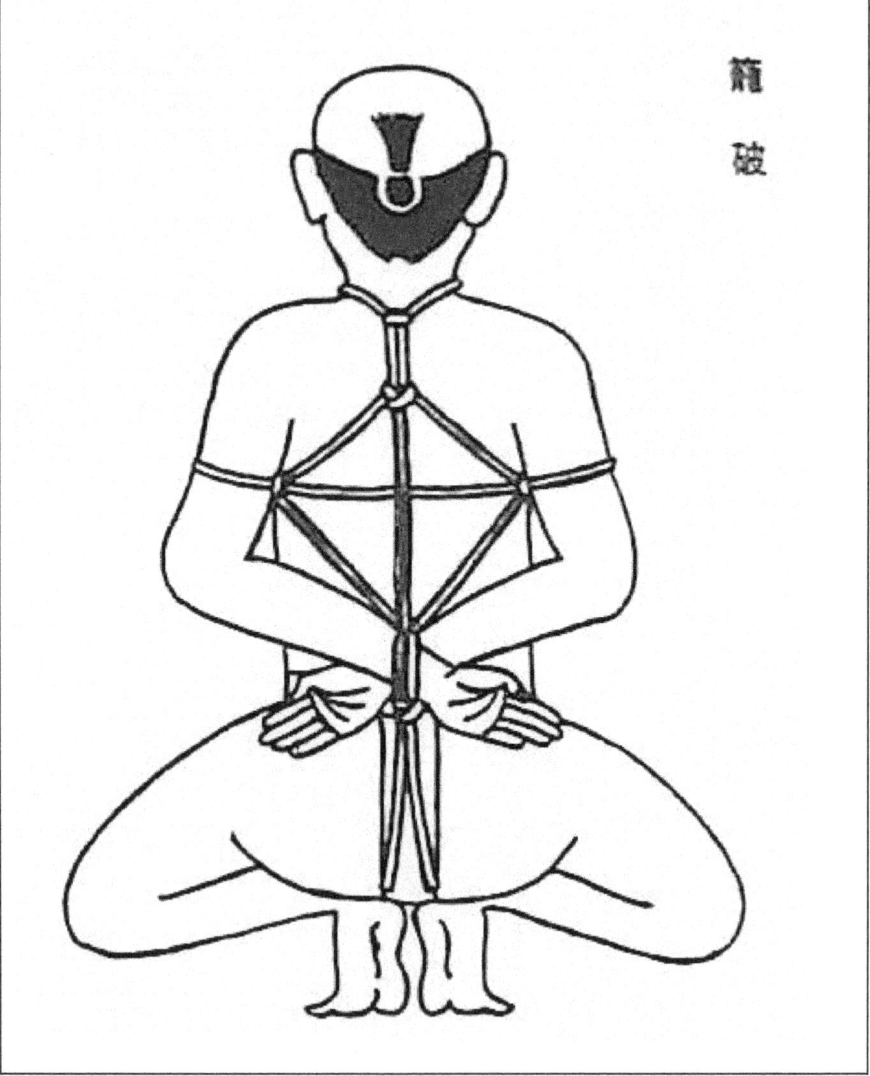

Igai "Wild Boar Valley" School
Senchu: Whilst on a Boat Whilst on a Boat is a straightforward tie. There is a Kuden about tying one wrist in front and one in back.
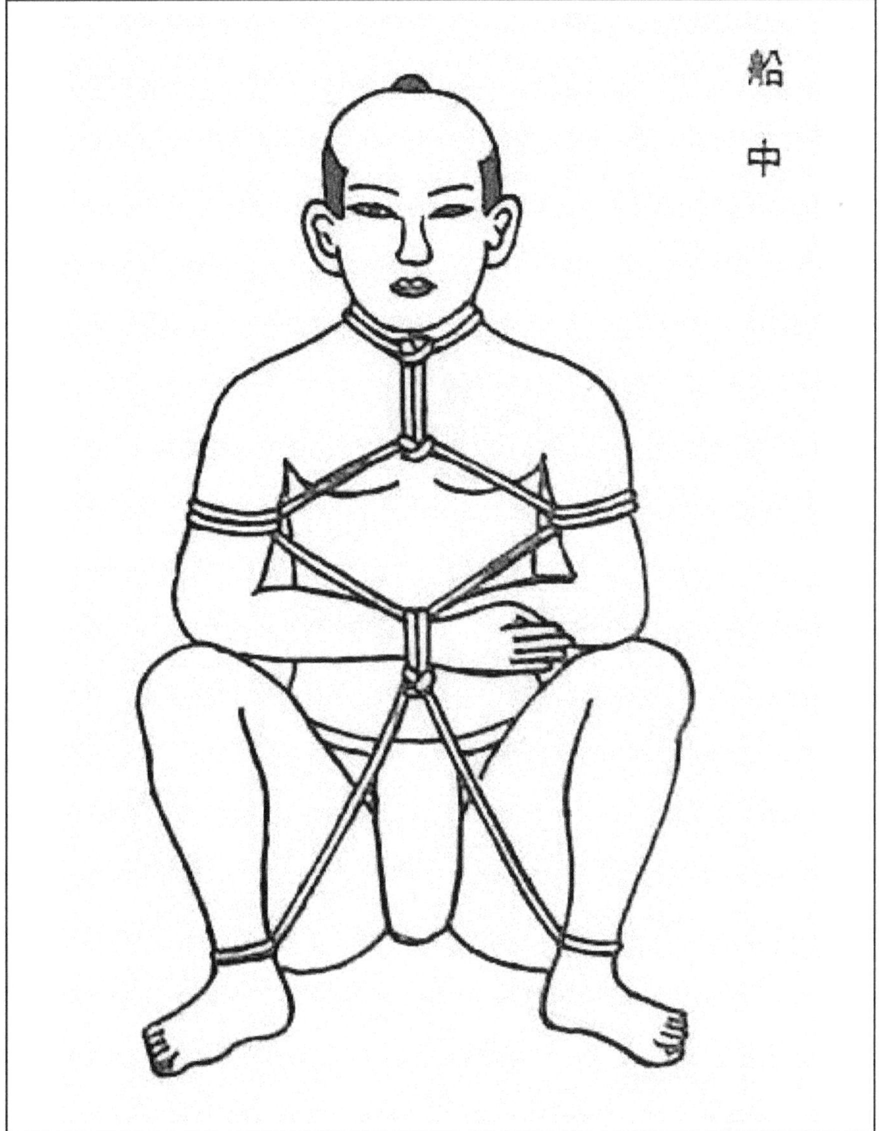

Igai "Wild Boar Valley" School

Mijin: Pulverizer

This is a tie used over the clothing and armor of a Samurai. The rope in this tie represents the Hachi Yo, the Seven Luminaries surrounding a single star, Kuji, the Nine Seals, Juji, the ten seals and Jukai, the Ten Buddhist Commandments.

Two strands of bowstring are used for this tie. The first tie, known as Purple Rope, is actually a bow string. There is a teaching about how to tie the rope. If near a Shinto shrine a Shime rope is used. If it is done before a temple the net and Horo on the back of a Samurai warrior will represent the womb and afterbirth. A Yin-Yang Knot is tied. There are many Kuden about East, West, South and North.

Note: Hachi Yo 八曜

 Image of the Eight Luminaries	Hachi Yo: This depicts the Seven Luminaries surrounding a central star, representing the person. The Seven Luminaries are the five visible planets plus the sun and moon. Mars (Fire) Mercury (Water) Jupiter (Wood) Venus (Metal) Saturn (Earth) This symbol was used as a family crest by Samurai families such as the Hosokawa Family.

Note: Shime

Shime is the rope used in Shinto Shrines to indicate an area has been purified. Below are some of the shapes a Shime rope can take.

前垂注連 Mae Dare Jime: Apron Rope Shaped

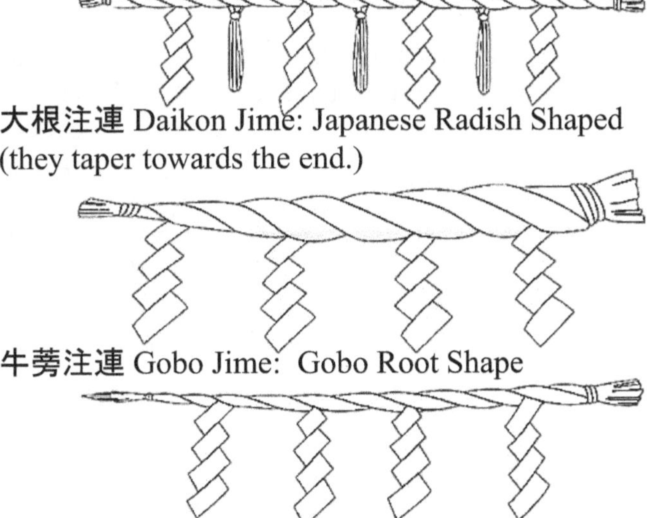

大根注連 Daikon Jime: Japanese Radish Shaped (they taper towards the end.)

牛蒡注連 Gobo Jime: Gobo Root Shape

Note: Kuji 九字

Zen	Zai	Retsu	Jin	Kai	Sha	Toh	Hyo	Rin

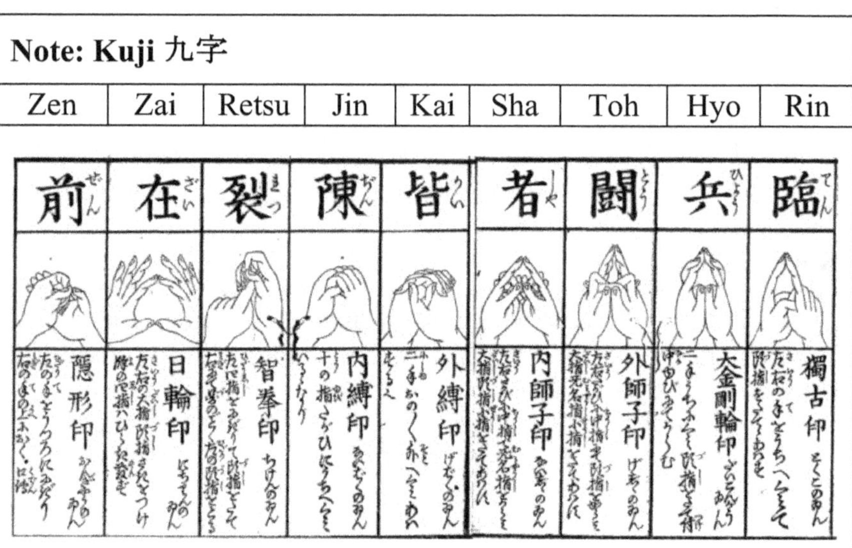

The Kuji (above, read from right to left) are Nine Hand Seals that are accompanied by invocations used in Shugendo and other ascetic traditions. An additional incantation and hand seal can be added at the end making Ju-ji, Ten Hand Seals.

Note: Horo 母衣 Arrow Cape

A Horo is a special cloak worn on the back of a mounted Samurai that inflated when riding at speed. The purpose was to protect against arrows shot at the back. There may have been a bamboo or whalebone frame underneath to keep the Horo spread open. A Samurai wearing a Horo on his back would have been part of a special unit. This unit that may have gone out past the vanguard to conduct strikes. The 16th century warlord Oda Nobunaga, who unified Japan, had two units of Samurai outfitted with Horo. The Black Horo Unit and the Red Horo Unit 黒母衣衆と赤母衣衆 each unit was comprised of 10 mounted soldiers. Interestingly, there was a general prohibition of displaying the severed head of a Samurai wearing a Horo. The reason for not displaying the severed head, known as Gokumon "Gates of Hell" 獄門 or Sarashi Kubi "Exposed Head" 晒首 is because the head will not be able to come into the light of the Buddha's grace. This deference to a Samurai wearing a Horo actually begins after the head is taken. It is wrapped in the Horo and handled reverently.

Illustrations: *Research into Horo* 保侶衣推考 by Ise Sadatake 伊勢貞丈 (1717-1784)

Additional Information: rekijin.com

Note: Jikkai 十戒 The Ten Buddhist Commandments
Jikkai are the Ten Commandments of Buddhist monks and nuns. They are also known as Guzoku Kai 具足戒. Monks are known as Biku 比丘 or Shami 沙弥 and nuns are Bikuni 比丘尼 or Sami-ni 沙弥尼.
Jikkai 十戒
1. 不殺生 (ふせっしょう) Do not kill any living thing 2. 不偸盗 (ふちゅうとう) Do not steal 3. 不淫 (ふじゃいん) Sex outside of marriage is forbidden 4. 不妄語 (ふもうご) Do not lie 5. 不飲酒 (ふおんじゅ) Do not drink alcohol Note: The first five are what all Buddhist practitioners should follow and are known as the Gokai 五戒. The following five are added for monks and nuns. 6. 不塗飾香鬘(ふずじきこうまん) Do not decorate or perfume the body or hair 7. 不歌舞観聴(ふかぶかんちょう) Do not watch singing or dancing 8. 不坐高広大牀 (ふざこうこうだいしょう) Do not sleep in a bed that is wide or higher than your knee 9. 不非時食 (ふひじしき) Do not eat between meals (two meals a day) 10. 不蓄金銀宝 (ふちくこんごんほう) Possess no wealth or valuables.

Buei Ryu

武衛流

Hobaku Zu

捕縛図

Illustrated Arresting Techniques of the Buei School of Martial Arts

Nawa no Shidai: List of Rope Techniques

縄之次第

早縄　くはへ縄右の手へ移し首へ廻し小手斗縛る　大事の囚人には早くはり縄をかくるなり　口伝

御法　さし合の縄なり　寸法七寸五分　侍にても下人にても苦しからず　侍は首に紙を巻くべし　口伝

落花　籠より引出す時荒縄也　首の根は引通し又切る所にて掛様口伝有　侍は首に紙を巻くべし　口伝

千鳥　下郎の縄なり　寸法七寸　首の付根より結び二ッ小手にも結びあり　口伝

村雲　法師又は女人などに掛る　寸法八寸高手に口伝　大事の如くうとにはわり縄を掛留二ッ　口伝

六道　侍を縛る　寸法六寸四方高手にかける手首に紙を巻き　二つには神前又は親兄の礼有るは結び
に紙を付べし　いたずら者には割縄を掛る留め二つ　口伝

微塵　免の縄なり　結び九つあり　九品の浄土を表す　又九字を結びこめるなり　神前にて志るしの
縄とも云へり　鎧武者　母衣武者いずれも同前　但し首に紙を幣にして四方に付　弓の弦三尺
三寸留めにかける結び二つ何れも侍は首に紙を巻き　弓の弦にて留めてしぼる　口伝

十文字　詰籠者の縄也　寸法四寸四方ひしの内にたつ縄有　侍は首又は小手にても紙を巻くべし　口伝

船中縄　前にて縛る　小手の縄左右のひざへかけ　小手内へ出し留るなり　口伝

制剛流縄目録

早縄　蜘之掛　四海羽返　胴縄　竹縄　下緒

本縄　五法　落花　千鳥　十文字　村雲　六道　山嵐　船中　微塵

<table>
<tr><td colspan="4">Buei Ryu Hobaku-zu
Arresting Techniques of the Buei School of Martial Arts</td></tr>
<tr><td colspan="4">Buei Ryu Hobaku Zu:
Illustrated Arresting Techniques of the Buei School of Martial Arts</td></tr>
<tr><td rowspan="4">武衛流縄縛図
武衛流は制剛流より出た梶原流より伝承の流派である。
指合（御法）
村雲
微塵
落花
籠破
十文字
千鳥
六道
船中</td><td colspan="3">This school is derived from the Kajiwara School, which is, in turn, derived from the Seigo School.</td></tr>
<tr><td>Mijin</td><td>Mura Kumo</td><td>Sashi Ai/Goho</td></tr>
<tr><td>Jumonji</td><td>Kago Yaburi</td><td>Rakka</td></tr>
<tr><td>Senchu</td><td>Rokudo</td><td>Chidori</td></tr>
</table>

Buei School	
	**Goho: Revered Way** This is also known as Sashi Ai no Nawa, Command Tie. The interval between knots is 7 Sun 5 Bun (22.5 cm.) This technique can be applied to both Samurai as well as servants. If done on a Samurai a piece of paper should be wrapped around his neck. There is a Kuden. Note: This technique is called Goho, but the Kanji are different from the other schools. It seems to resemble the technique by the same name from the Igai School but is far less complex than the one in the Seigo School.
	**Rakka: Fallen Flower** Fallen Flower is tied on people after they have been taken out of a palanquin. Use rough cord. There is a Kuden regarding how the rope should pass around the base of the neck as well as how the knot over the place on the neck where the executioner will cut should be tied. If done on a Samurai a piece of paper should be wrapped around his neck. There are other Kuden as well.

Buei School	
	**Chidori: Plover** This is used on servants. The interval between ties should be 7 Sun (21 cm.) Two knots should be tied at the base of the neck. There is a tie at the wrists as well. Kuden.
	**Mura Kumo: Cloudy Village** This tie is used for Buddhist Monks, womenfolk and the like. The interval between ties is 8 Sun (24 cm.) There is a Kuden about tying the upper arms. It is imperative that the Wari Nawa, Split Knot, is tied correctly according to the time of year. Kuden.

Buei School	
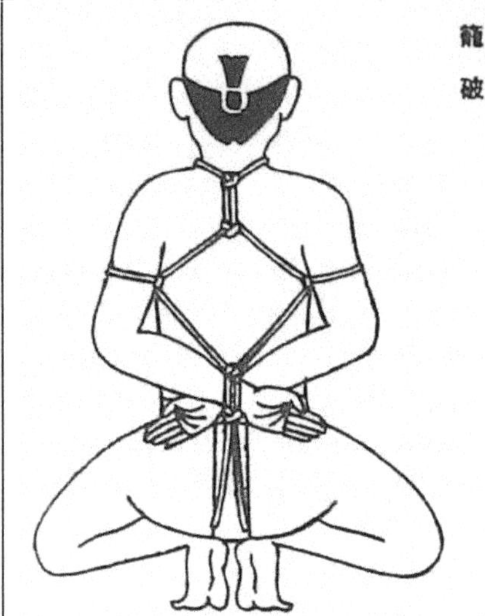	**Kago Yaburi** [No description]
	**Rokudo: Six Paths** This tie is used to restrain Samurai. The interval between ties should be 6 Sun (18 cm.) Paper should be wrapped around all points on the upper arms and the wrists. There are two other situations that call for wrapping paper around a body part before tying a knot. The first is when you are in front of a Shinto Shrine or Buddhist Temple. The other if the parents or brother are present as you are applying the rope. This is a show of respect. If the prisoner is behaving in a troublesome manner, then a Wari Nawa, Split Knot, should be tied with two knots at the bottom. Kuden.

Buei School

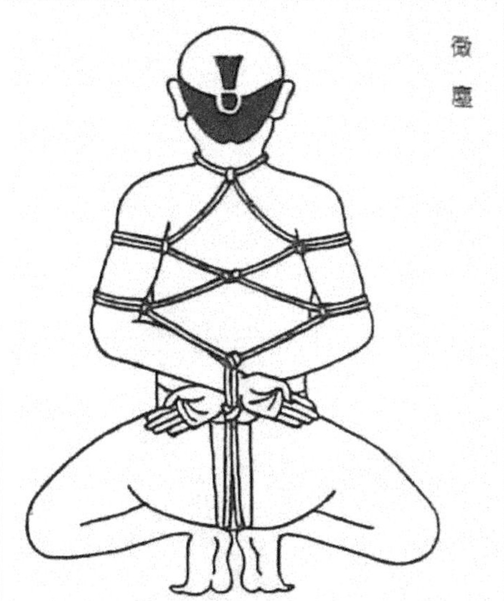

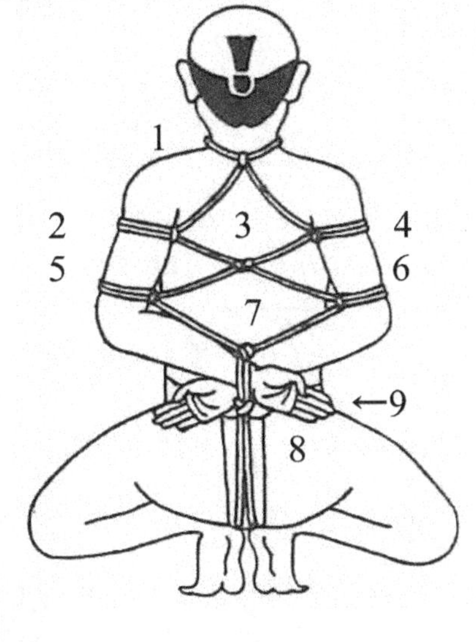

Mijin: Pulverizer

This is a banishment or removal from office tie. There are a total of nine ties that represent the Nine Amida of Rebirth. They also represent the Kuji, the Nine Binding Seals. In addition, the rope is said to represent the will of the gods.
This tie can be used to restrain both Samurai in armor as well as those with a Horo on their backs. However, paper is placed around the neck along with the 4 Directions, representing the upper arms and wrists. The tie is with a 3 Shaku 3 Sun (99 cm) bowstring with two knots at the end. No matter the rank, all Samurai should have paper wrapped around their neck. The bow string should be wound tight at the end. Kuden.

Note: I presume the 9th tie is on the front of the Fundoshi, loin cloth.

Note: Amida Kuhon-in 阿弥陀九品印
The Nine Levels of Amida Rebirth
Amida was possibly a monk named Dharmakāra who achieved enlightenment or Buddhahood. Devotees of this Buddhism are born into one of the nine Pure Lands. Pure Land is composed of nine different levels or grades. The illustrations below show each level with a different Mudra or hand gesture.
Illustrations: Butsuzo-zu-i 仏像図彙 *Illustrated Collection of Buddhist Images* 1690 by Tosa Hidenobu 土佐秀信.

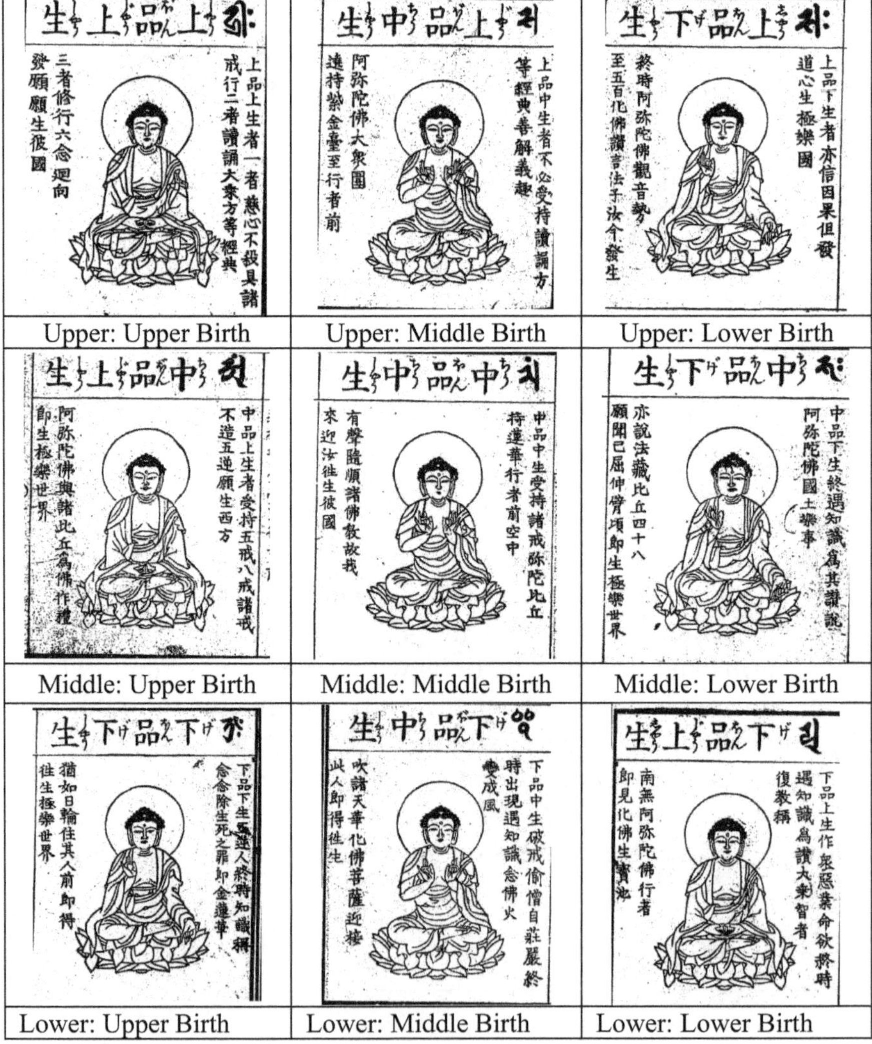

Upper: Upper Birth	Upper: Middle Birth	Upper: Lower Birth
Middle: Upper Birth	Middle: Middle Birth	Middle: Lower Birth
Lower: Upper Birth	Lower: Middle Birth	Lower: Lower Birth

Buei School	
 	**Jumonji: Cross Shaped** This tie is used for a person being carried in a palanquin. The interval between all the ties should be 4 Sun (12 cm.) A Tachi Nawa, Standing Knot, should be tied at all points. If the prisoner is a Samurai, then paper should be wrapped around the neck and wrists.
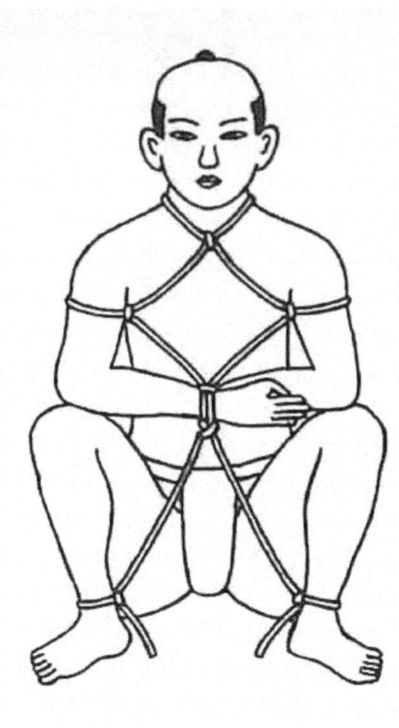	**Senchu Nawa: Whilst on a Boat** This is tied on the front of the prisoner. When tying the wrists, the left and right arms should be placed on the knees. They should be tied off, so they stay in place. Kuden. Note: This sounds like the person is bound up in a tripod with their butt and two feet forming the base. The reason might be to add stability against the rocking of a boat, though Japanese craft were often travelling across rivers more than across open ocean.

Rigoku Ryu

理極流

Hobaku Zu

捕縄術

Bondage Techniques of the Rigoku School Of Martial Arts

Rigoku Ryu **Techniques from the Rigoku School**			
五方 十文字 微塵	微塵 Mijin Pulverizer	十文字 Jumonji Cross Shaped	五方 Goho Five Directions
落花 村雲		村雲 Murakumo Cloudy Village	落花 Rakka Fallen Flower
千鳥 六道		六道 Rokudo Six Paths	千鳥 Chidori Plover

Goho: Five Directions	Rakka: Fallen Flower
五方	落花

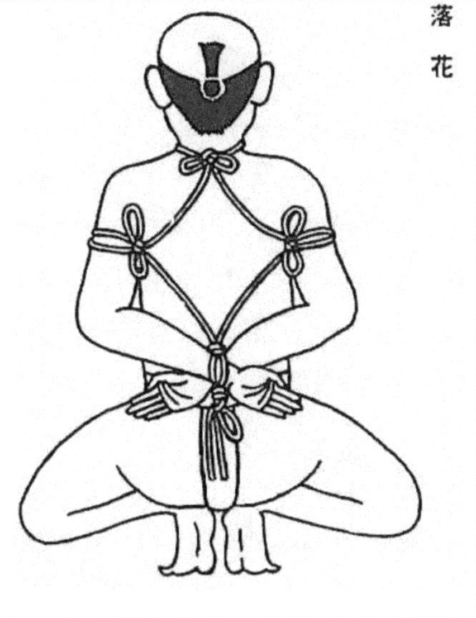

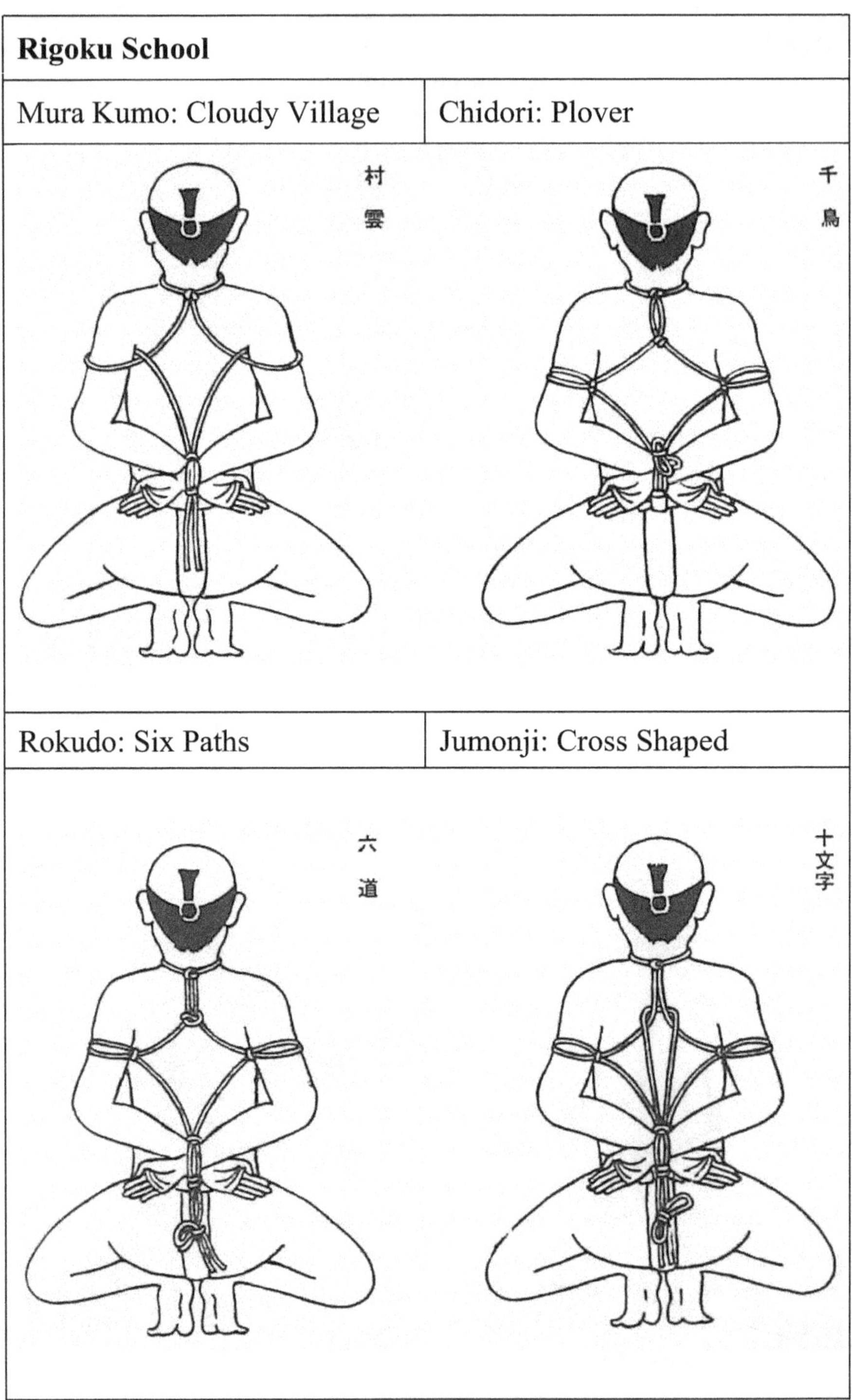
Rigoku School
Mura Kumo: Cloudy Village
Chidori: Plover
村雲
千鳥
Rokudo: Six Paths
Jumonji: Cross Shaped
六道
十文字

Rigoku School

Mijin: Pulverizer

Bondage Techniques from the:

Araki Ryu   Seishin Ryu

荒木流・清心流

Shingoku Ryu   Joshin Ryu

心極流・常慎流

Muso Ryu

夢想流

Schools of Martial Arts

Araki School・Seishin School・Shingoku School

These teachings, which originated in the Shingoku School are practiced in the Araki School and the Seishin School.

Kuden.

Note: Kuden is the first word in this document but it is not clear if this means the following all contain Kuden, there is an overall Kuden at the beginning or that these are Kuden that happened to be written down.

荒木流清心流心極流

本流は心極流より出て荒木流、清心流と伝りたるもの。

口伝

一 早縄 図の通り

一 五方 五ヶ所（首 胴 高手 両方の小手）ヲ戒シムルユヘ五方ト云フ

但シ 真 行 草アリ

真ハ 菱 二ッ 胴縄 二ッ 行ハ 菱 一ッ 胴縄 一ッ 草 胴縄 無シ菱斗リ

一 十文字 強力ノ者ニ掛ル縄 十文字ニ成故十文字と云フ 小手ニテ三ケ縄タルミナシ

一 村雲 ムラ〳〵ト出ル雲 サラリトキヱル理ナリ

一 落花 命ヲワルト花のチルコトシ 切縄ナリ

一 位 上総(アゲマキ)具足ノ背中ニ下リタル緒ニテ首小手折返シニフサ引通ス大将ノ類等縛ル故ニ位ト云フ

一 微塵 縄ノアマリ引ト塵ニシマル故ナリ

一 早縄 早縄ハ右ノ小手ヲククリ首ヘ廻シ左ノ小手ヲトリ両小手ノ上ヲカラム

一 五法 五方ハ首ヘ一重廻シヒシヲ結ヒ高手二ッ廻シ小手ノトコロ男結其下同断但シフシノ間三寸

一 十文字 十文字ハ強力者ニカケル縄ナリ 初メ五方ノ通リニカケ小手ノ余リタル縄ヲ高手ヘ通シヒシノ所ニテ男結ニスル

一 落花 落花ハ首ニテ男結ニシテ高手ヲ上ヨリ廻シハサケ小手男結上下首ヲトケハ高手モトケル也

一 村雲 小手男結上下

Araki School・Seishin School・Shingoku School

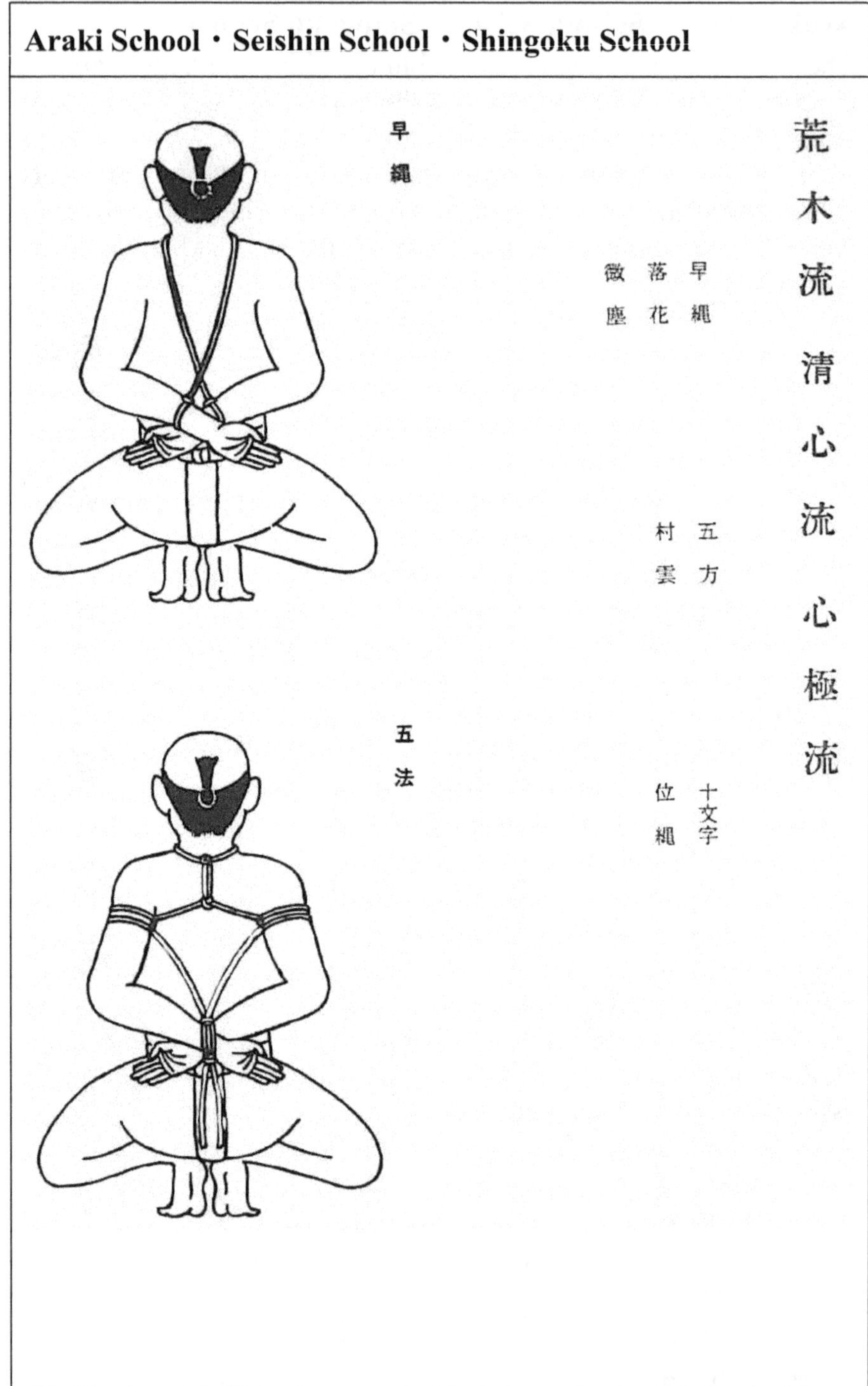

Araki School・Seishin School・Shingoku School These teachings, which originated in the Shingoku School are practiced in the Araki School and the Seishin School.			
早縄 落花 微塵	微塵 Mijin Pulverizer	落花 Rakka Fallen Flower	早縄 Haya Nawa Fast Tie
五方 村雲		村雲 Murakumo Cloudy Village	五方 Goho Five Directions
十文字 位縄		位縄 Kurai Nawa	十文字 Jumonji Cross Shaped

<table>
<tr><th colspan="2">Araki School・Seishin School・Shingoku School</th></tr>
<tr><td></td><td>Haya Nawa: Fast Tie

This is done as shown in the illustration.</td></tr>
<tr><td></td><td>Goho: Five Directions
This tie is done on five areas: Neck, Torso, Upper Arm and both Wrists. Since there are five ties on five different points on the body it is called Five Directions. However, there are three levels of this with differing complexity. This is the Shin-Gyo-So, True-In Between-Free Application.
Shin, the true or fundamental way to tie this knot has 2 Water Chestnut Knots along with 2 Waist Knots.
Gyo, In-Between, is halfway between the fundamental way and the following Free Application. Free Application has 1 Water Chestnut Knot and 1 Waist Knot.</td></tr>
</table>

<table>
<tr><th colspan="2">Araki School・Seishin School・Shingoku School</th></tr>
<tr><td>
</td><td>Jumonji: Cross Shaped Tie

This is a tie used on strong prisoners. When completed the tie resembles the Kanji Ju 十, representing the number 10 as well as a cross shape, thus it is called Cross Shaped Tie. The wrists should have three ties, without any slack in the rope.</td></tr>
<tr><td>
</td><td>Rakka: Fallen Flower

The meaning of this tie is that your life is broken just as a flower blossom falls from a tree. This is a Kiri Nawa, a tie used for execution by decapitation.</td></tr>
</table>

Araki School・Seishin School・Shingoku School

Mura Kumo: Cloudy Village

This tie is done with a "Mura-Mura" sound, which describes a wide layer of clouds approaching and, just like clouds, they disperse and fade away. This is the feeling of this technique.

Note:

So, after multiple versions of Mura Kumo appearing in this book the meaning behind the name finally becomes clear. The name is more accurately translated as *A Wide Swath of Cloud Rolling in and then Thinning out and Dissipating*. There are a great many more onomatopoeia in Japanese than in English and they can convey a wider variety of sensations. For example, *Uji-Uji* is the "sound" of a person hesitating and *Bura-Bura* is the "sound" of something swinging back and forth or of wandering aimlessly.

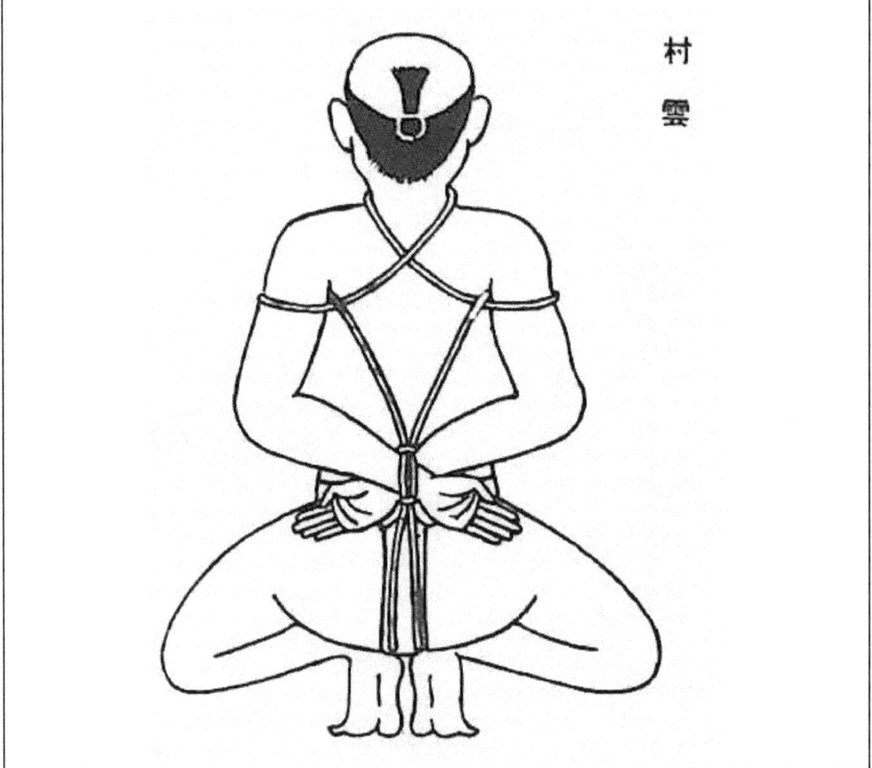

Araki School・Seishin School・Shingoku School

Kurai Nawa:
Status Tie

A tie that resembles the four points of the compass. It is tied on the back of the armor, descending downward. The knot around the neck and hands fold down to the next section. This is called Kurai, Status Tie, because it is used to secure Taisho, or Generals.

Mijin: Pulverizer

The remaining rope is pulled through the knot that is why the last Kanji means "leftover" or "trash."

Note on Mijin: The word Mijin means to pulverize into tiny fragments, however in this case the meaning is "using the remaining rope to continue the tie" so the same word Mijin could be translated as Remainder or Scrap of Rope as interpreted by this school.

Araki School・Seishin School・Shingoku School
Note: These techniques have no illustrations.
Haya Nawa: Fast Tie To apply this Fast Tie, first tie the rope around the right wrist then wrap it once around the neck. Take hold of the left wrist and tie the hands on top of each other.
Goho (With different Kanji): Five Methods To apply Five Methods, wrap the rope once around the neck and tie a Water chestnut Knot. Two wraps should be done around the upper arms and a Men's Knot shout tie off the hands. The rest is the same, however the interval between ties should be 3 Sun (9 cm.) Note: "The rest is the same" might be referring to the technique of the same name above. From that point on tie the Five Methods the same way as Five Directions.
Jumonji: Cross Shaped Cross Shaped is a rope technique to restrain strong people. First tie Goho, Five Directions. With the rope left over from tying the wrists, secure the upper arms and tie Men's knots at the elbows.
Rakka: Fallen Flower When tying Fallen Flower, a Mens's Knot should be tied at the neck. Avoid tying the upper arms from above, the wrists are tied with a Men's Knot. When untying the upper neck and lower neck knots the upper arms should be untied as well. Note: This technique is hard to understand, possibly intentionally.
Mura Kumo: Cloudy Village A Men's Knot should be tied both above and below the wrists.

Joshin School ・Muso School
Note: There are no descriptions for the techniques

常慎流夢想流

小手返
小手縄

早縄
芝居縄

マワシ縄
早縄問

小手返

早縄

Joshin School ・Muso School	
小手縄 Kote Nawa Wrist Rope	小手返 Kote Gaeshi Reversed Hand
芝居縄 Shibai Nawa Stage Play Tie	早縄 Haya Nawa Fast Tie
早縄間 Haya Nawa Aida Intermediate Fast Tie	マワシ縄 Mawashi Nawa Twist of Rope

Haya Nawa: Fast Tie	Kote Gaeshi: Reversed Hand

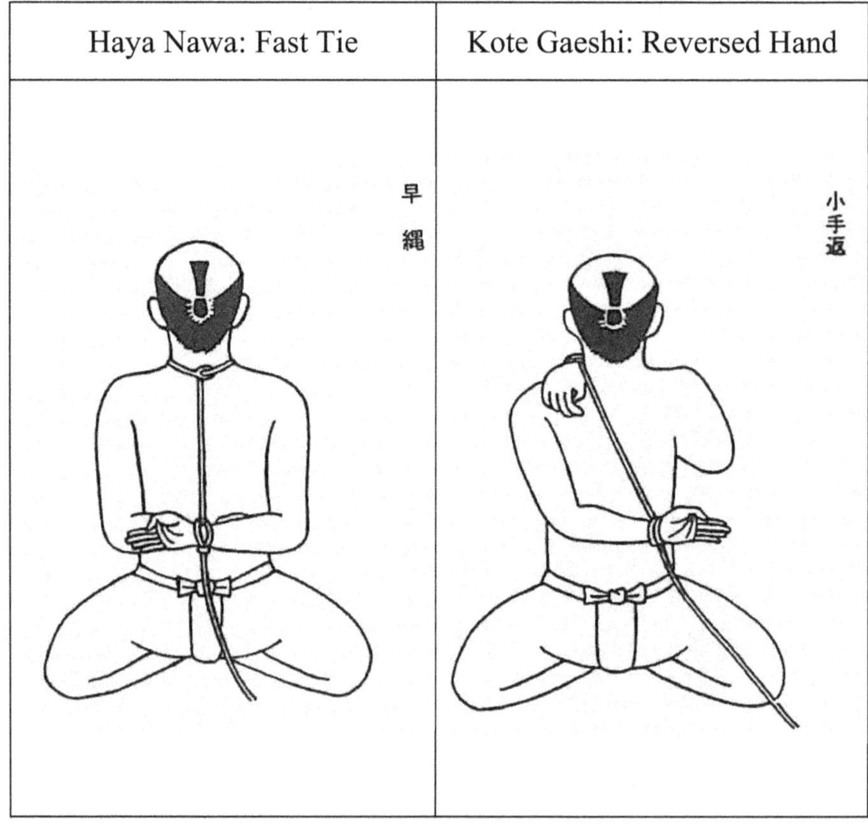

Joshin School ・Muso School	
Shibai Nawa: Stage Play Tie	Mawashi Nawa: Twist of Rope
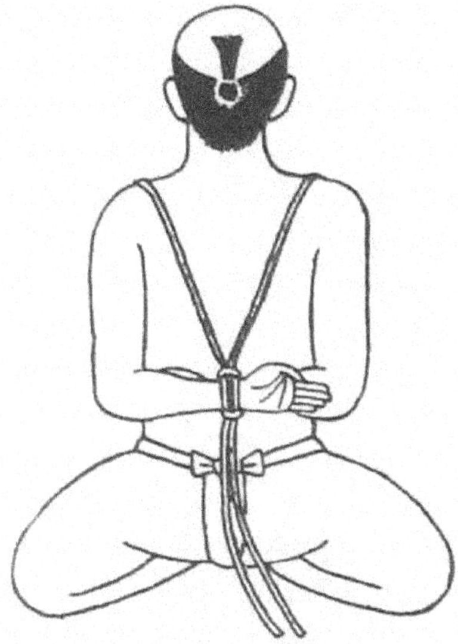	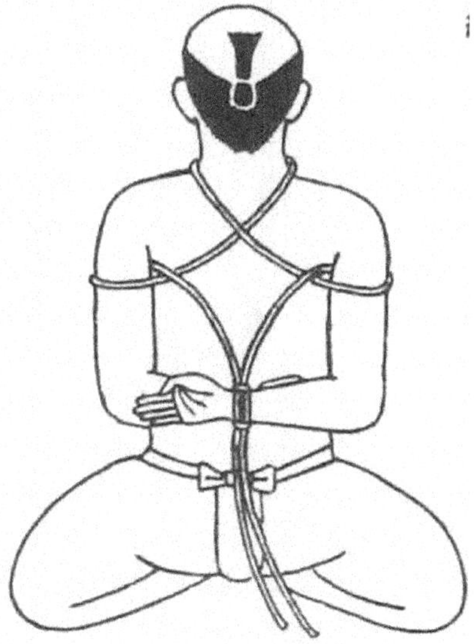
Haya Nawa Aida: Intermediate Fast Tie	Kote Nawa: Wrist Tie
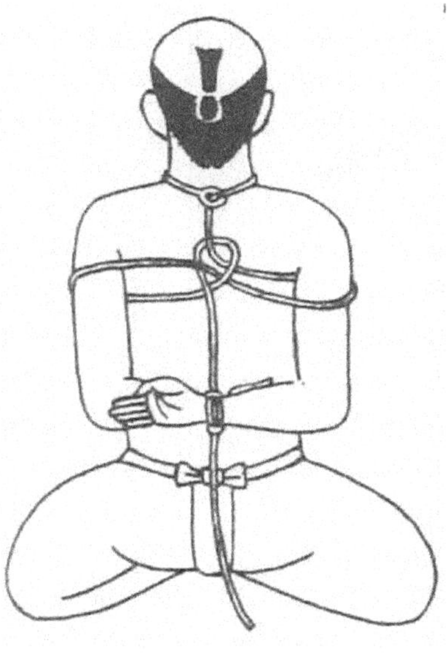	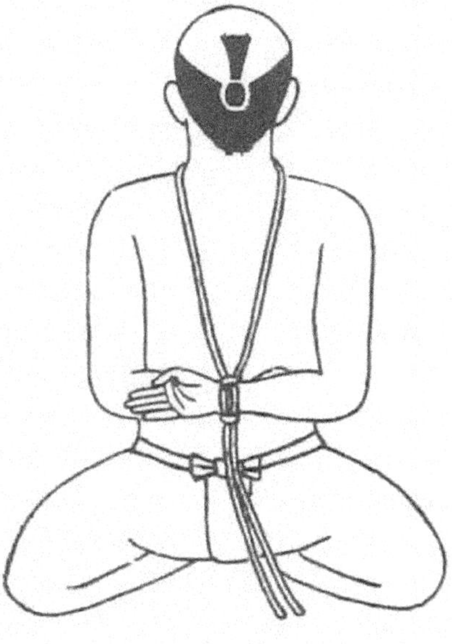

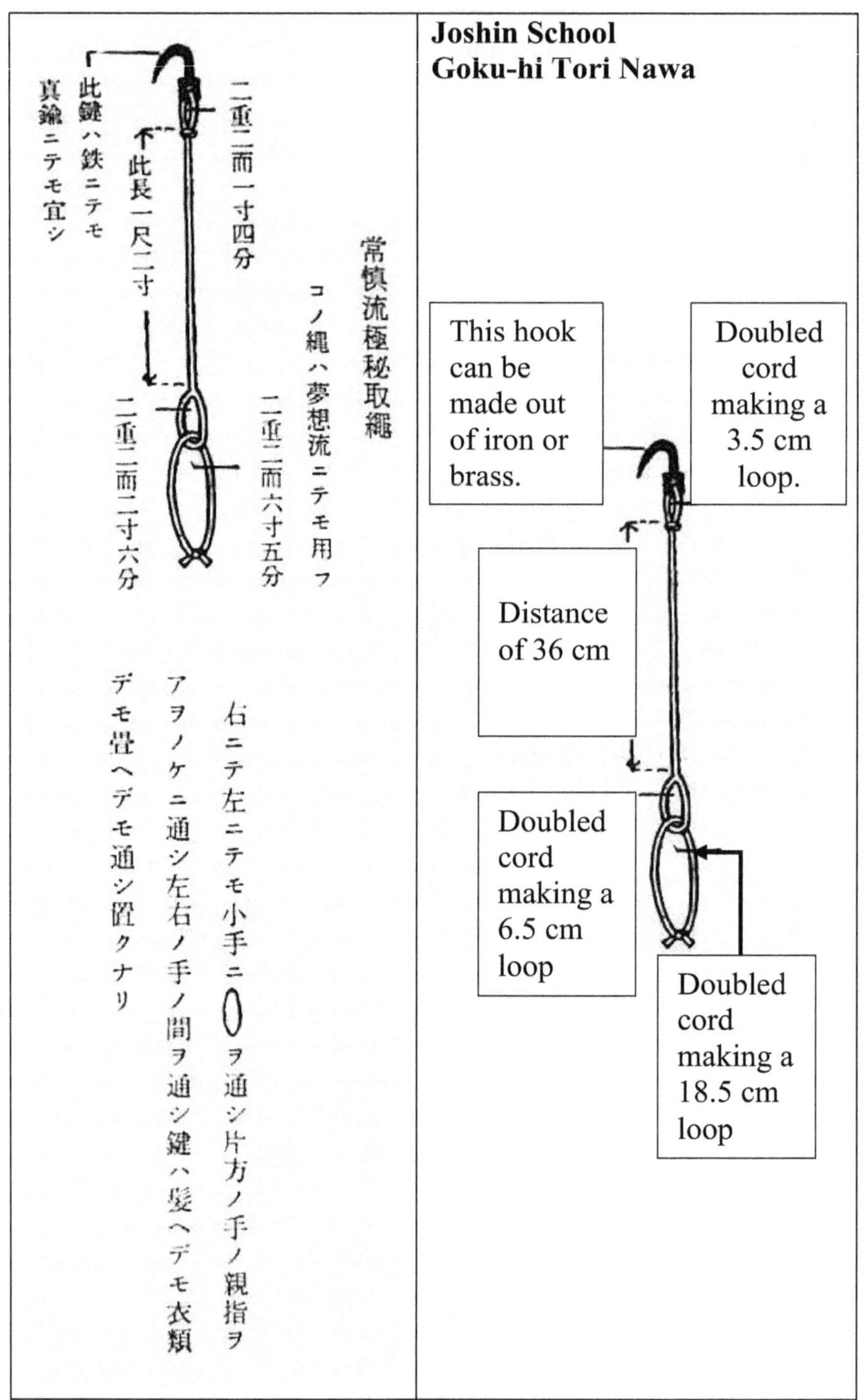
常慎流極秘取縄
コノ縄ハ夢想流ニテモ用フ
二重二而六寸五分
二重二而一寸四分
此長一尺二寸
二重二而二寸六分
此鍵ハ鉄ニテモ
真鍮ニテモ宜シ
右ニテ左ニテモ小手ニヲ通シ片方ノ手ノ親指ヲ
アヲノケニ通シ左右ノ手ノ間ヲ通シ鍵ハ髪ヘデモ衣類
デモ畳ヘデモ通シ置クナリ
Joshin School
Goku-hi Tori Nawa
This hook can be made out of iron or brass.
Doubled cord making a 3.5 cm loop.
Distance of 36 cm
Doubled cord making a 6.5 cm loop
Doubled cord making a 18.5 cm loop

Joshin School
Goku-hi Tori Nawa
Inner Secret Arresting Technique
This technique is also used in the Muso School

右ニテ左ニテモ小手ニハ O ヲ通シ
片方ノ手ノ親指ヲアヲノケニ通シ
左右ノ手ノ間ヲ通シ鍵ハ髪ヘデモ衣類
デモ畳ヘデモ通ツ置クナリ

Pull either the prisoner's left or right hand through the O shaped loop at the bottom. Take the thumb of the other hand and hold it upright and pass it through the loop on top of the other hand. Wrap the cord between the two hands. The hook can be attached to either the hair, clothing or a Tatami mat.

Note: The instructions are pretty spartan, so this is my best interpretation. The instructions mention left and right hands a lot and it is tough to even determine if the instructions refer to the person doing the technique or the prisoner. Overall, this tool seems to be used to rapidly and firmly restrain a violent person.

Nanba Ippo Ryu

難波一甫流

Bondage Techniques of the Nanba Ippo School of Martial Arts

難波一甫流 Nanba Ippo Ryu
Note: These techniques contain no descriptions

難波一甫流

七曜
番不入
括索
六道縄
禁縄

火責
不動加縦縛
手組縄
早縄
手擾索

二重菱
胴搦
真之胴
天狗縄
口伝

難波一甫流 Nanba Ippo Ryu	
七曜	Shichi Yoh: Seven Luminaries
火責	Hizeme: Torture By Fire
二重菱	Niju Hishi: Double Water Chestnut
番不入	Ban Irazu: No Need For A Guard
不動加羅縛	Fudo Karabaku: Restrained With Lord Fudo's Lariat
胴搦	Do Karame: Tangled Waist
括索	Kassaku: Capture Tie
手組縄	Tegumi Nawa: Hand Capture Tie
真之胴	Shin no Do: True Waist Tie
六道縄	Rokudo Nawa: Six Paths Tie
早縄	Haya Nawa: Fast Tie
天狗縄	Tengu Nawa: Mountain Goblin Tie
禁縄	Kin Nawa: Forbidden Tie
手繦索	Te Mutsuki Saku: Hand Diaper Tie
口伝	Kuden: Orally Transmitted Technique

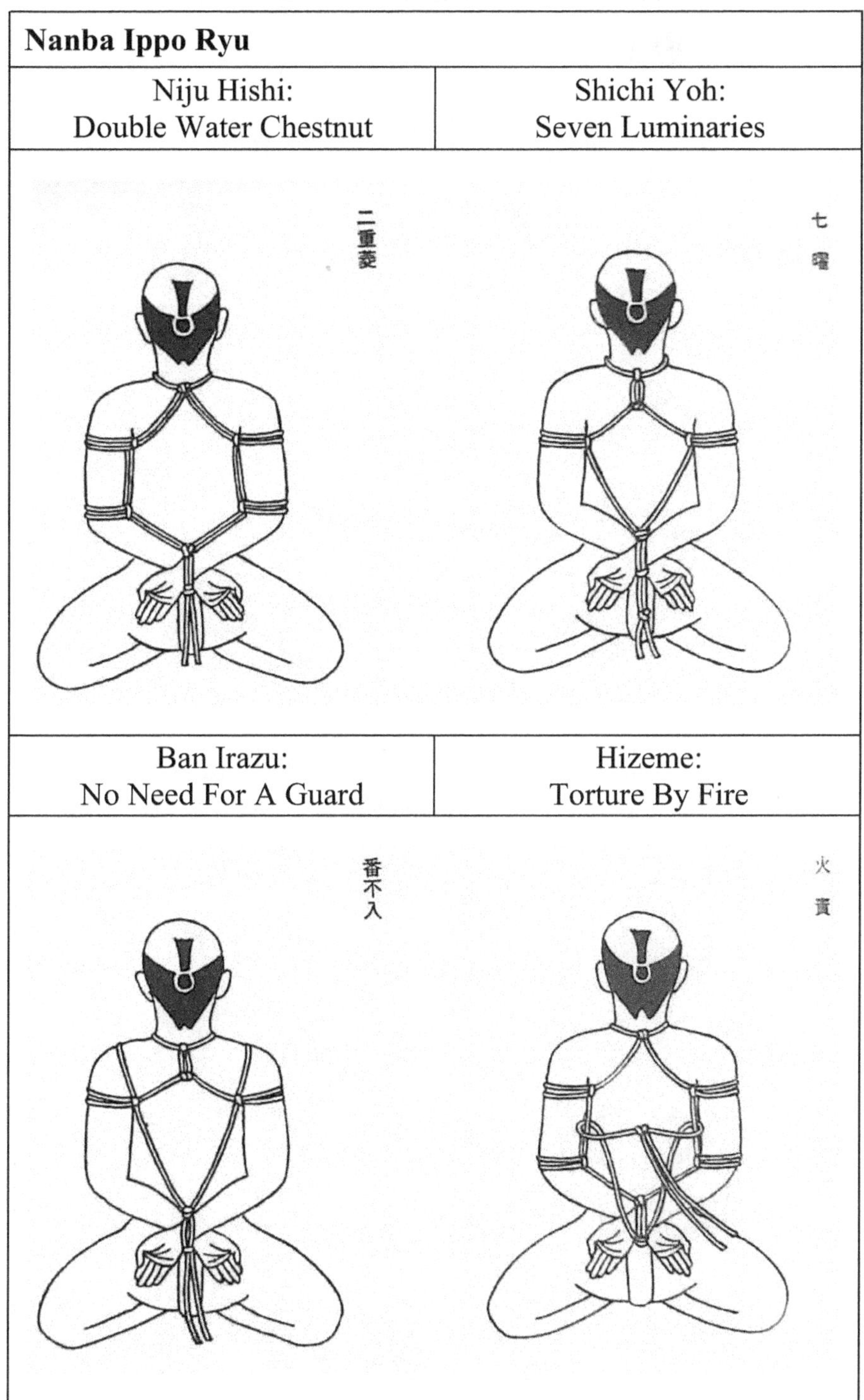

Nanba Ippo Ryu

Niju Hishi: Double Water Chestnut	Shichi Yoh: Seven Luminaries
Ban Irazu: No Need For A Guard	Hizeme: Torture By Fire

Nanba Ippo Ryu	
Do Karame: Tangled Waist Front View	Fudo Karabaku: Restrained With Lord Fudo's Lariat. Back View
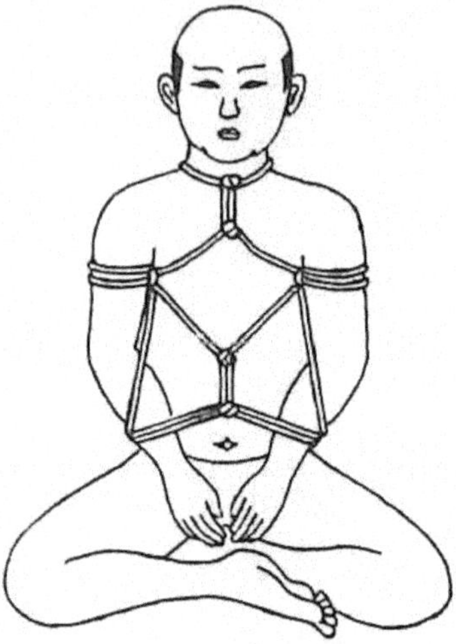	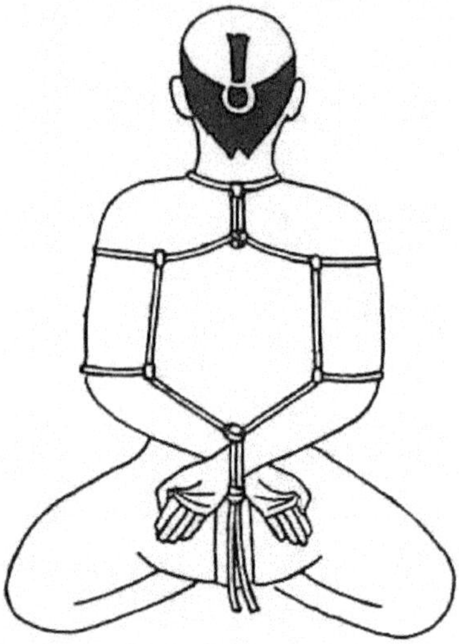
Do Karame: Tangled Waist Back View	Fudo Karabaku: Restrained With Lord Fudo's Lariat. Front View
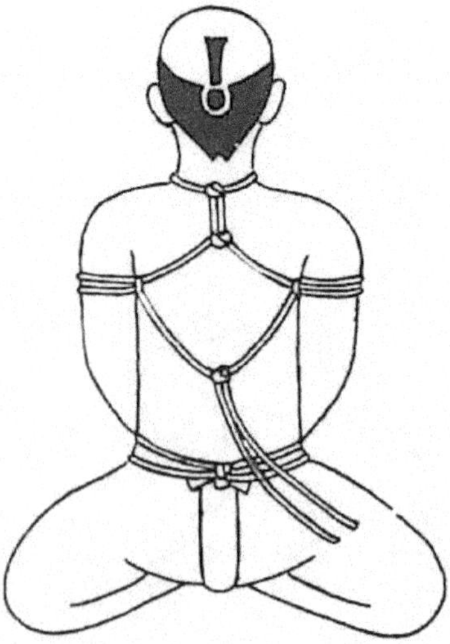	

Nanba Ippo Ryu	
Tegumi Nawa: Hand Capture Tie Back View	Kassaku: Capture Tie Front View
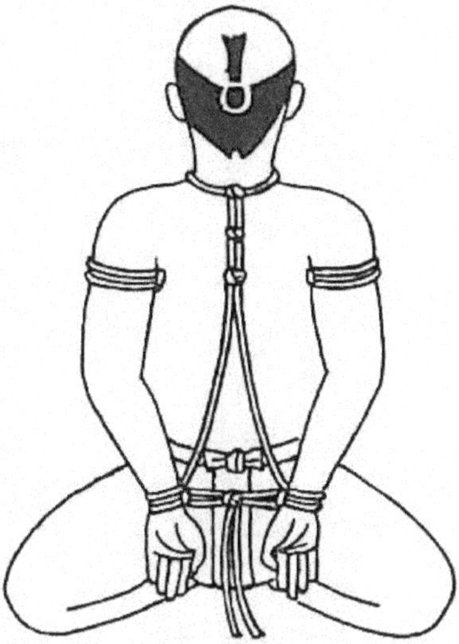	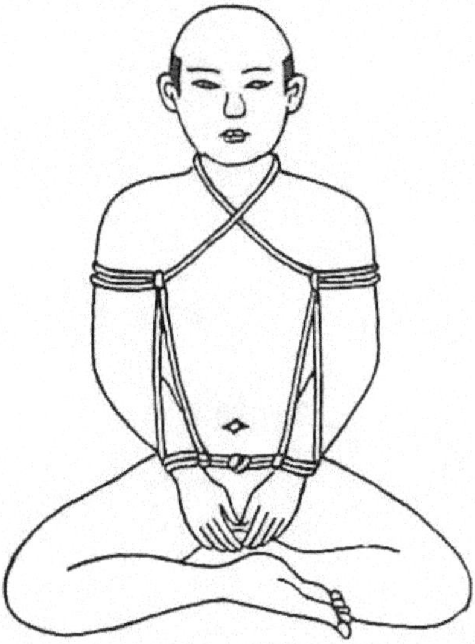
Tegumi Nawa: Hand Capture Tie Back View	Kassaku: Capture Tie Back View
	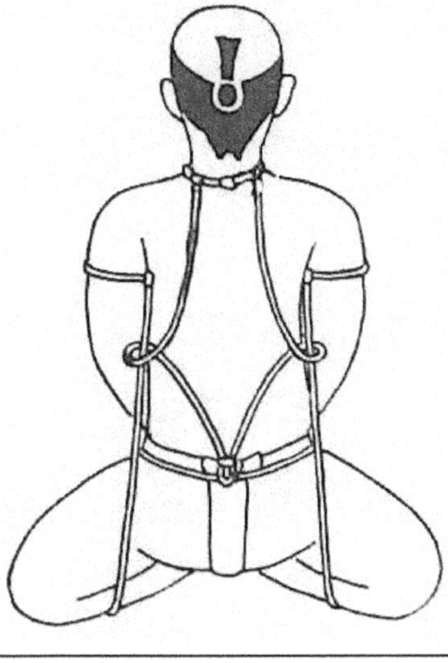

Nanba Ippo Ryu	
Roku Do Nawa: Six Paths Tie	Shin no Do: True Waist Tie Front View

Haya Nawa: Fast Tie	Shin no Do: True Waist Tie Back View

Nanba Ippo Ryu	
Te Mutsuki Saku: Hand Diaper Tie	Tengu Nawa: Mountain Goblin Tie
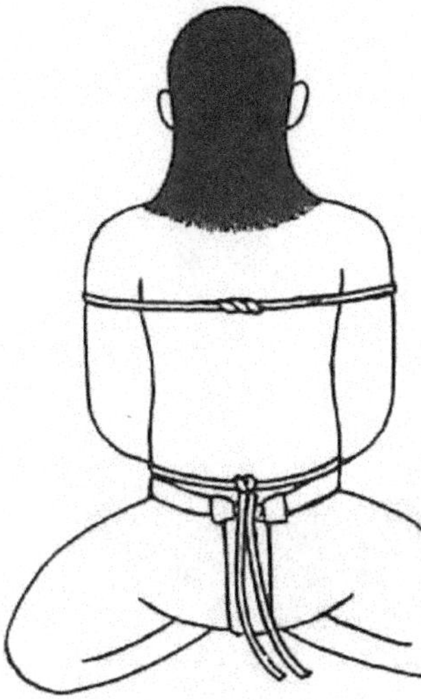	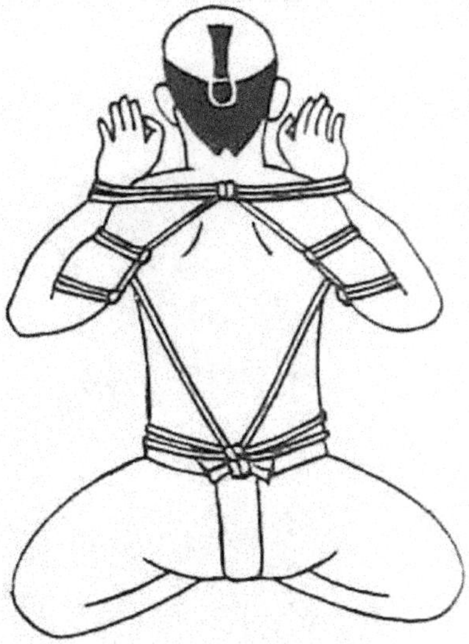
Kuden: Orally Transmitted Technique	Kin Nawa: Forbidden Tie
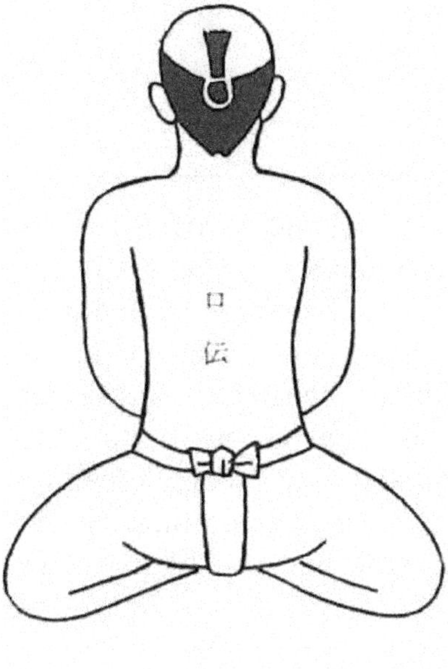	

Note: Tengu

Tengu 天狗 or Mountain Goblins are Japanese gods/ magical creatures inhabit the mountains and forests. The Kanji are "Heavenly Dog" though they can also have more avian features. The word Tengu comes from China but referred to an inauspicious celestial event. In Japan they began to be associated with supernatural tricksters who preyed on wayward Buddhist monks. Later, their imaged changed and they became creatures who could be sought out for mystical learning and martial arts. They are often shown wearing the garb and cap of mountain ascetics.

Miyamoto Musashi slicing the wings off a Tengu by Tsukioki Yoshitoshi 月岡芳年 (1839-1892)

Toh Ryu

東流

Bondage Techniques of the Toh "East" School of Martial Arts

東流 Toh Ryu

東流

火責

不動加羅縛

胴搦 後前

天狗縄

口伝

二重菱

手組索

活縄 後前

禁縄

番不入

真之胴

早縄

手繦縄

東流 Toh Ryu Note: These techniques contain no instructions.	
火責	Hizume: Torture By Fire
二重菱	Niju Hishi: Double Caltrop
番不入	Ban Irazu: No Need For a Guard
不動加羅縛	Fudo Karabaku: Restrained With Lord Fudo's Lariat
手組索	Tegumi Saku: Rope Handcuffs
真之胴	Shin no Doh: True Waist Tie
胴搦　後前	Doh Karame Go Zen: Waist Tangle Front and Back
活縄　後前	Katsu Nawa Go Zen: Resuscitating Tie Front and Back
早縄	Haya Nawa: Fast Tie
天狗縄	Tengu Nawa: Mountain Goblin Tie
禁縄	Kin Nawa: Forbidden Tie
手縕縄	Tegarami Nawa: Hand Tangle Tie
口伝	Kuden: Orally Transmitted Technique

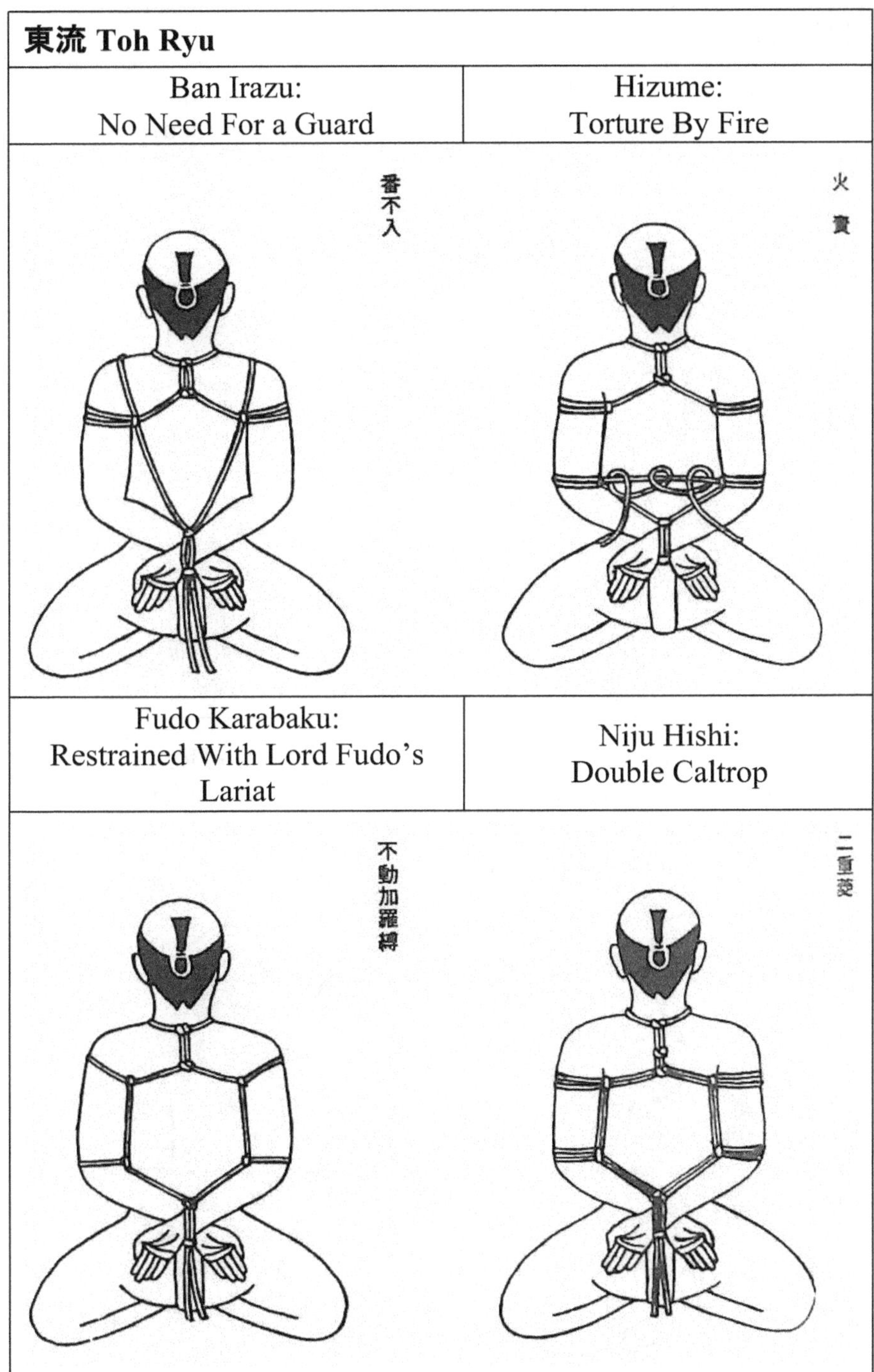

東流 Toh Ryu	
Ban Irazu: No Need For a Guard	Hizume: Torture By Fire
Fudo Karabaku: Restrained With Lord Fudo's Lariat	Niju Hishi: Double Caltrop

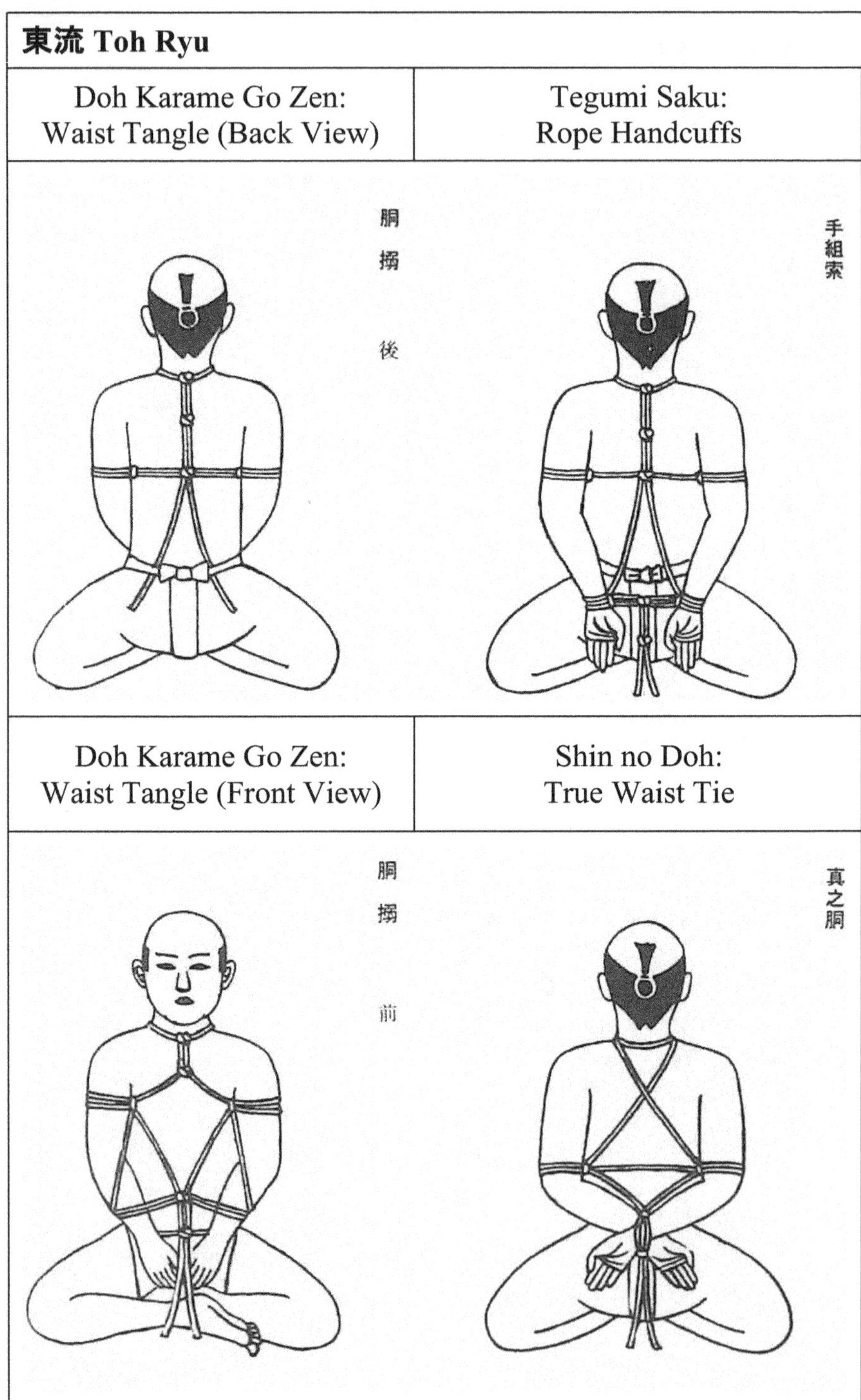
東流 Toh Ryu
Doh Karame Go Zen:
Waist Tangle (Back View)
胴搦後
Tegumi Saku:
Rope Handcuffs
手組索
Doh Karame Go Zen:
Waist Tangle (Front View)
胴搦前
Shin no Doh:
True Waist Tie
真之胴

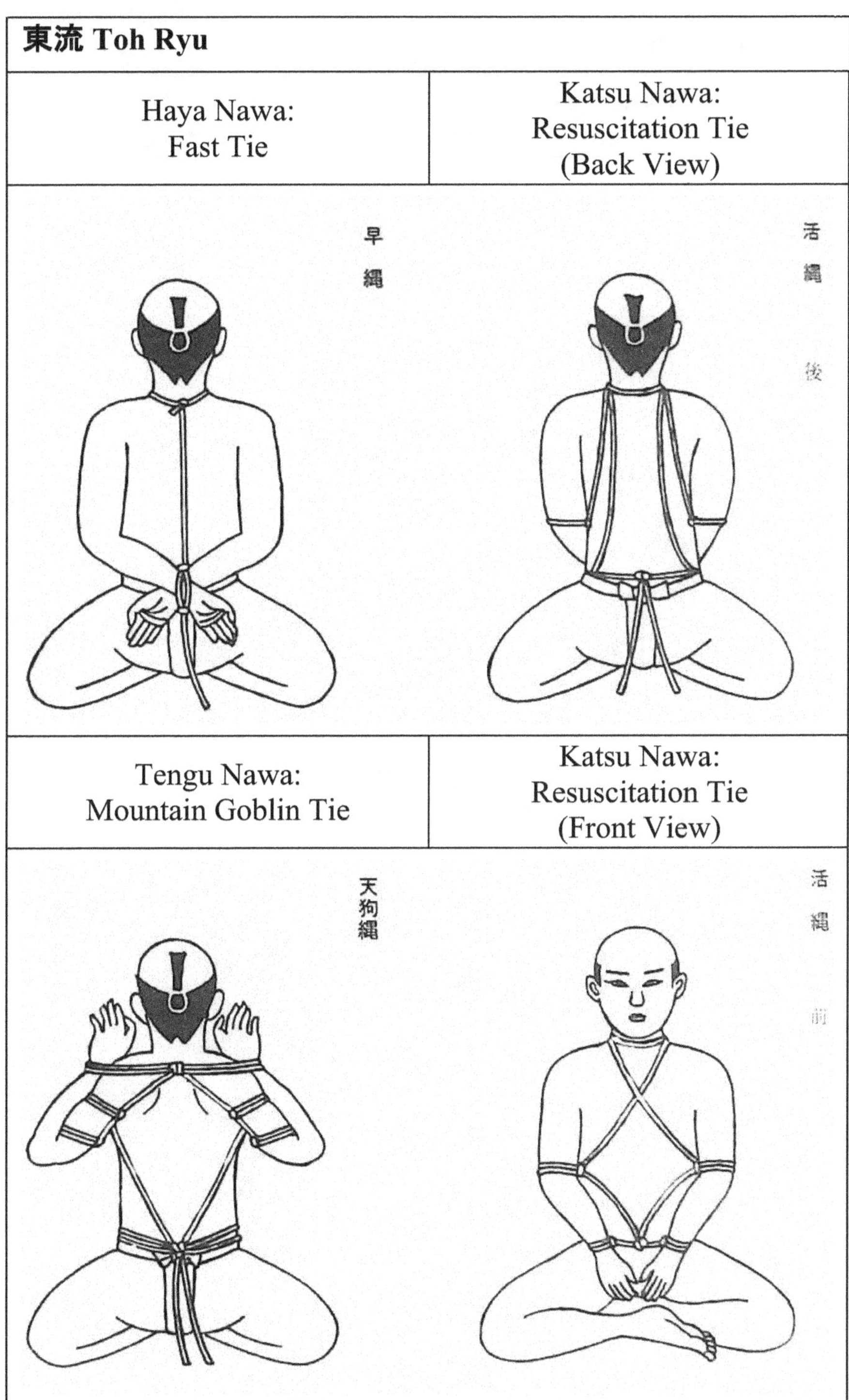
東流 Toh Ryu
Haya Nawa:
Fast Tie
Katsu Nawa:
Resuscitation Tie
(Back View)
早縄
活縄 後
Tengu Nawa:
Mountain Goblin Tie
Katsu Nawa:
Resuscitation Tie
(Front View)
天狗縄
活縄 前

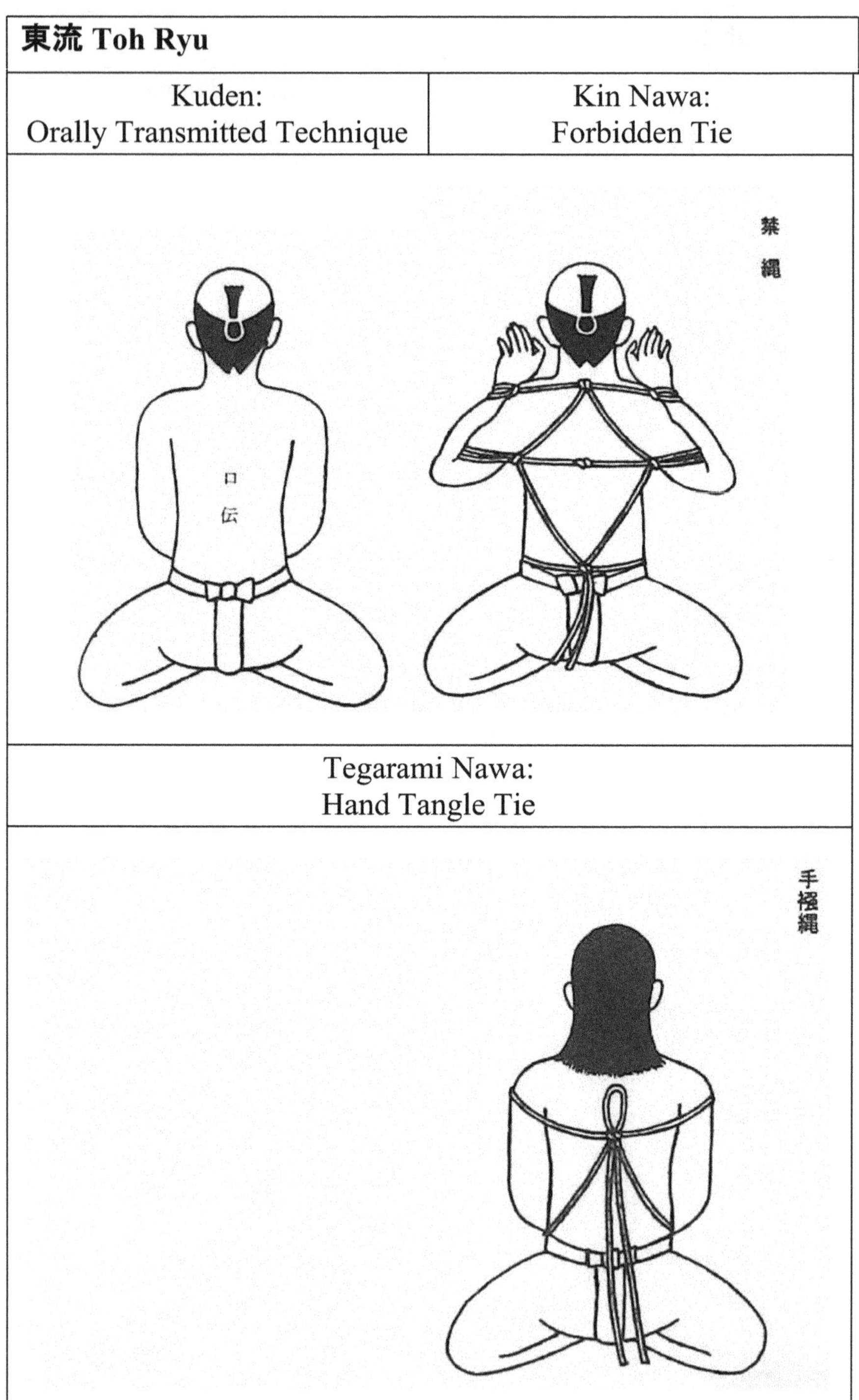
東流 Toh Ryu
Kuden:
Orally Transmitted Technique
Kin Nawa:
Forbidden Tie
禁縄
口伝
Tegarami Nawa:
Hand Tangle Tie
手搦縄

Nanba Ippo Ryu

難波一甫流

Toh Ryu

東流

Nawakake Hiden

縄掛秘伝

Secret Bondage Techniques of the Nanba Ippo & Toh Schools of Martial Arts

Secret Bondage Techniques of the Nanba Ippo & Toh Schools

難波一甫流　東流縄掛秘伝

急所　早縄之事　戒縄之事　早縄ニテ本縄ノ懸様之事

追放縄　留縄留様　八寸縄之事　小手付ノ事

ハガイメノ事　官僧正縄懸様留様　無官出家等縄掛様

Secret Bondage Techniques of the Nanba Ippo & Toh Schools	
急所	Kyusho: Vital Point
早縄之事	Haya Nawa: Fast Tie
戒縄之事	Kainawa no Koto: Concerning the Warning Rope
早縄ニテ本縄ノ懸様之事 Haya Nawa Nite Hon Nawa no Kake Yoh no Koto: How to Apply Main Tie Using Fast Tie	
追放縄	Tsuiho Nawa: Banishment Tie
留縄留様	Tome Nawa Tome Yoh: Stopping Rope For Stopping
八寸縄之事	Hassun Nawa no Koto: How to Use the 24 cm Rope
小手付ノ事	Kote Tsuke no Koto: On Restraining The Hands
ハガイメノ事	Hagai-jime no Koto: Double Arm Lock
官僧正縄懸様留様 Kan Zo Sho Nawa Kake Yoh Tome Yoh: Proper Arrest Tie for Abbots of Buddhist Temples	
無官出家等縄掛様 Mu Kan Shukke Toh Nawa Kake Yoh: Proper Arrest Tie for Non-Abbot Buddhists and Travelling Monks	

Secret Bondage Techniques of the Nanba Ippo & Toh Schools
The dots indicate where the rope is tied. There are three upper points, three central points and three lower points, for a total of 9 points. There are Kuden you must hear about how the legs and hands are tied. There are other Kuden about the illustrations at left.

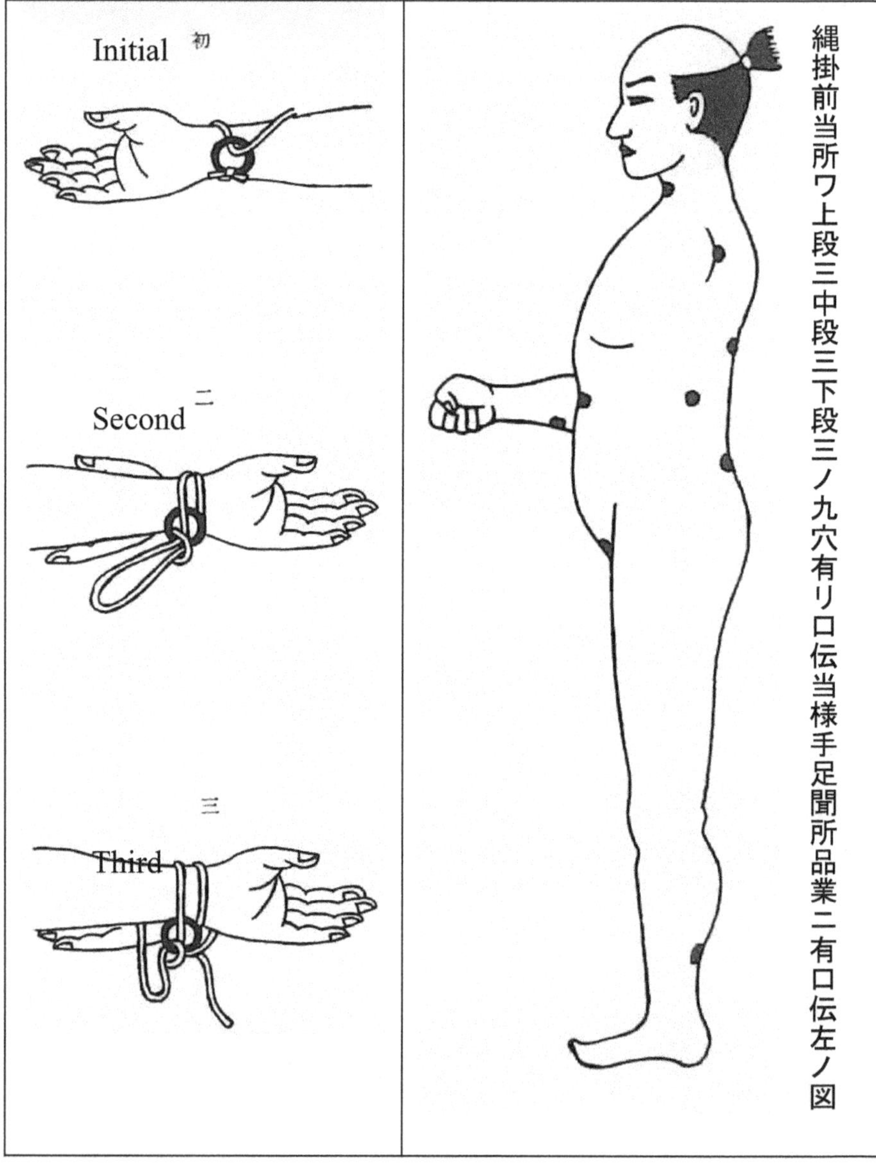

Secret Bondage Techniques of the Nanba Ippo & Toh Schools

Shidome: Four Way Stop
Kai Nawa no Koto: How to Apply the Warning Rope

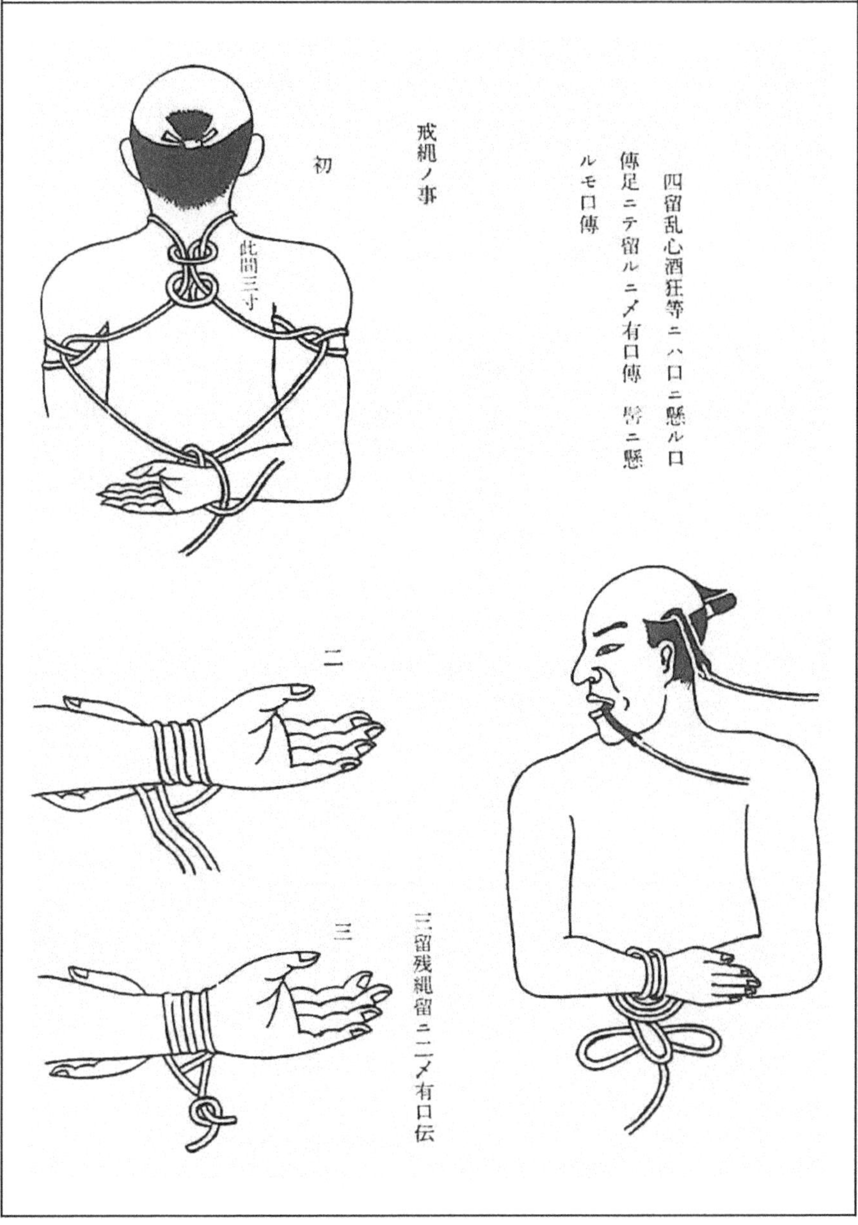

Secret Bondage Techniques of the Nanba Ippo & Toh Schools	
Shidome: Four Way Stop	
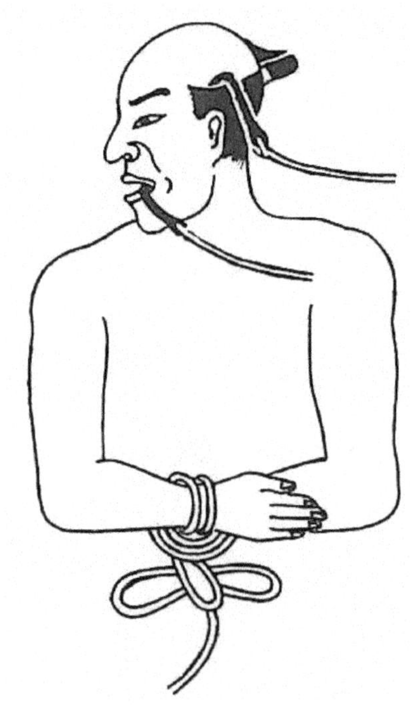	When restraining violent, mentally disturbed and extremely drunk people hook the iron hook in their cheek. Kuden. There is a tie for the feet. Kuden. You can also hook the hair. Kuden.

Secret Bondage Techniques of the Nanba Ippo & Toh Schools
Kai Nawa no Koto: How to Apply the Warning Rope Illustration 1: The interval between these knots should be 3 Sun (9 cm.)
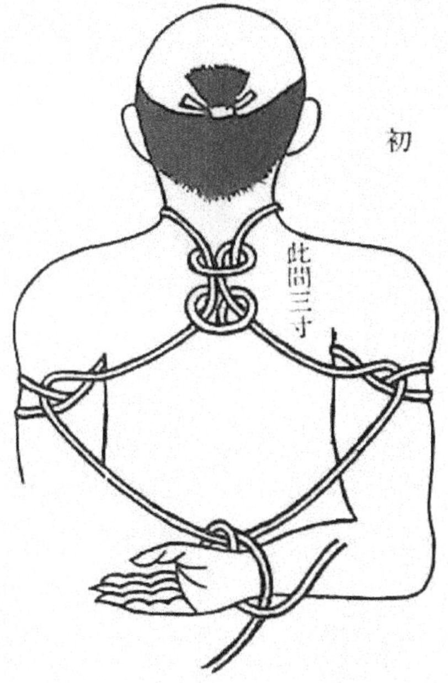
Illustration 2
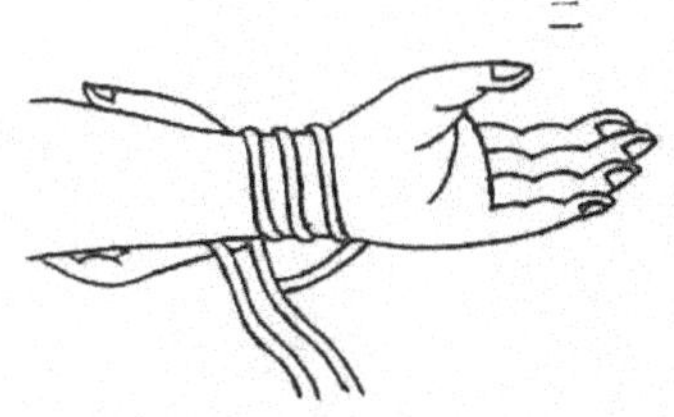
Illustration 3: How to tie off the remaining rope. Use two knots. Kuden.
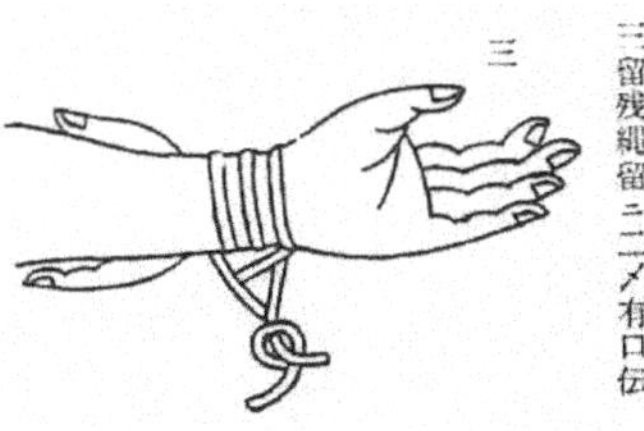

Secret Bondage Techniques of the Nanba Ippo & Toh Schools

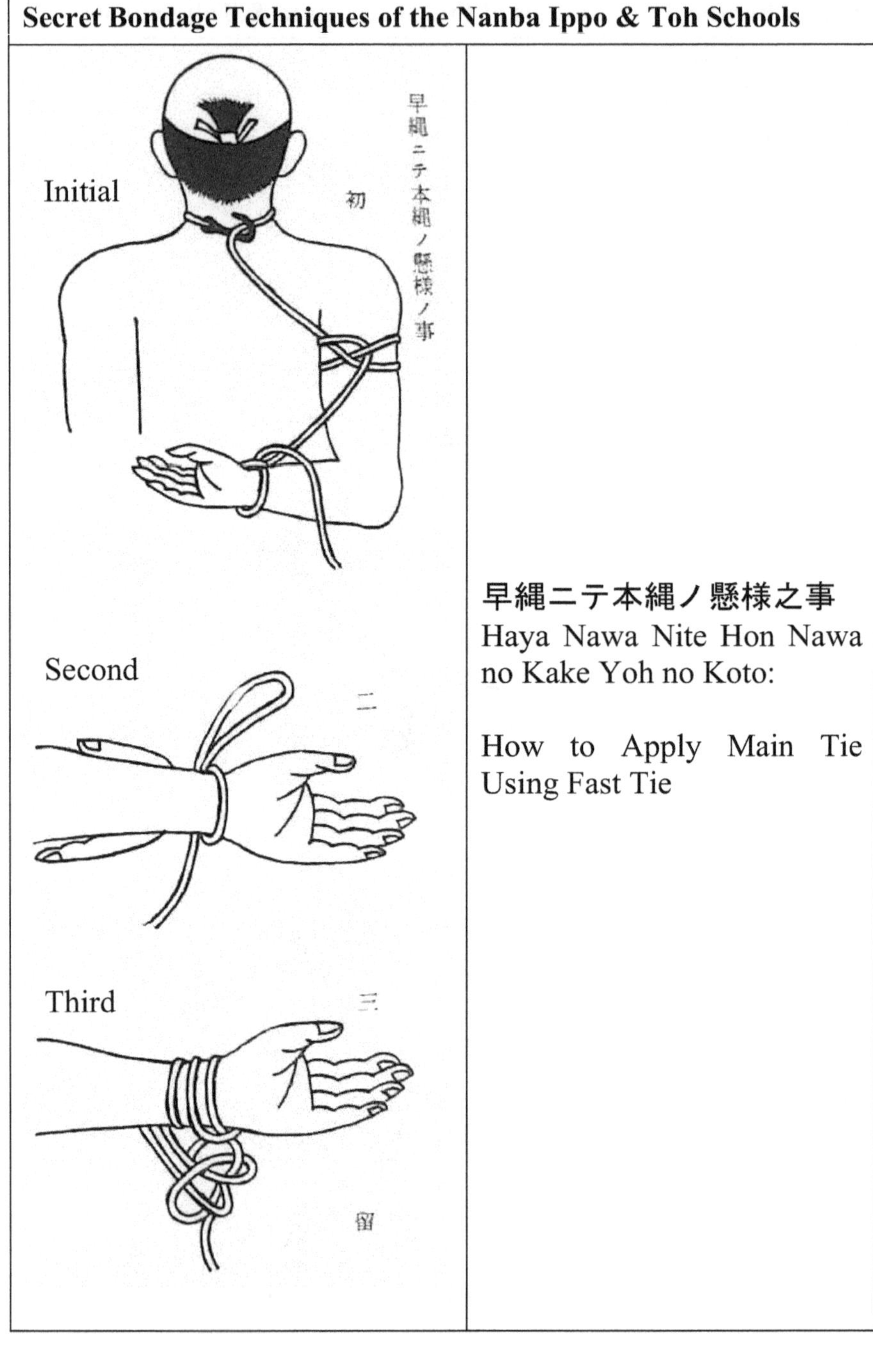

早縄ニテ本縄ノ懸様之事
Haya Nawa Nite Hon Nawa no Kake Yoh no Koto:

How to Apply Main Tie Using Fast Tie

Secret Bondage Techniques of the Nanba Ippo & Toh Schools	
追放縄楊枝ヲ 用事口傳楊枝無 キ時ワ針ヲ用ウ 口傳 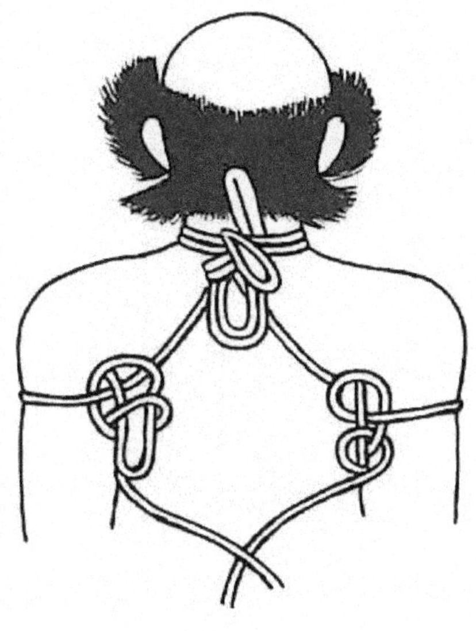	追放縄楊枝ヲ用事口傳楊枝無キ時ワ針ヲ用ウ口傳 Tsuiho Nawa: Banishment Tie There is a Kuden, an oral only transmission about using a Yohji 楊枝, toothpick (or small pointed piece of wood.) If you do not have a Yohji then a Hari, or needle, can be used. Kuden.
留縄留様戒縄 包無シ引ホトキ ニククル 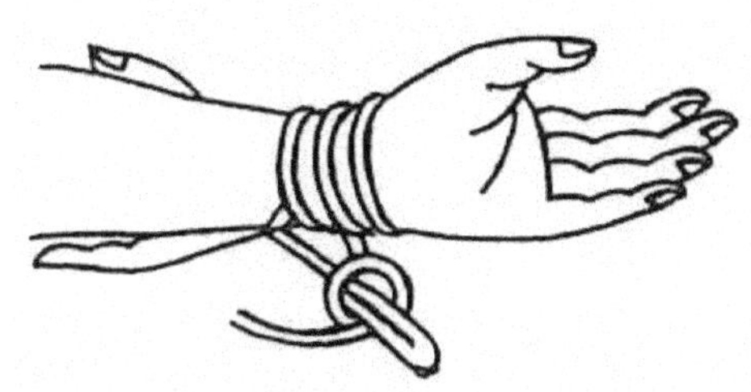	留縄留様戒縄 包無シ引ホトキニククル Tome Nawa Tome Yoh: Secure Stopping Tie Don’t wrap it, instead pull it through and twist. Note: These instructions are very vague.

Secret Bondage Techniques of the Nanba Ippo & Toh Schools

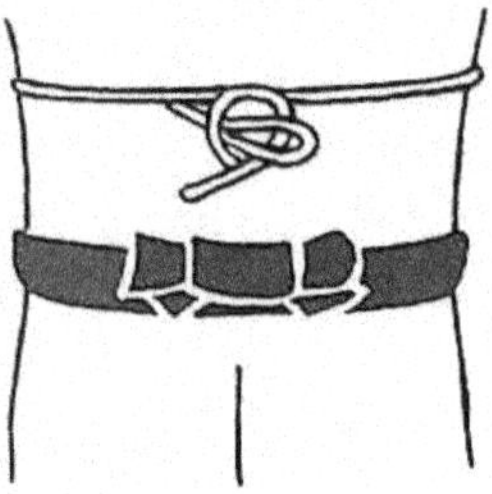

後留

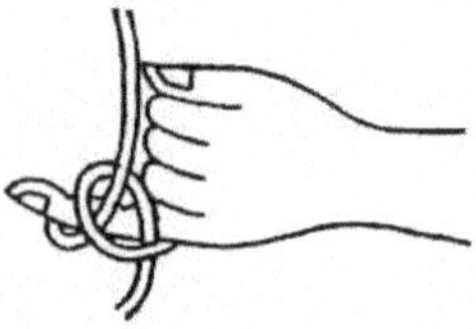

八寸縄之事

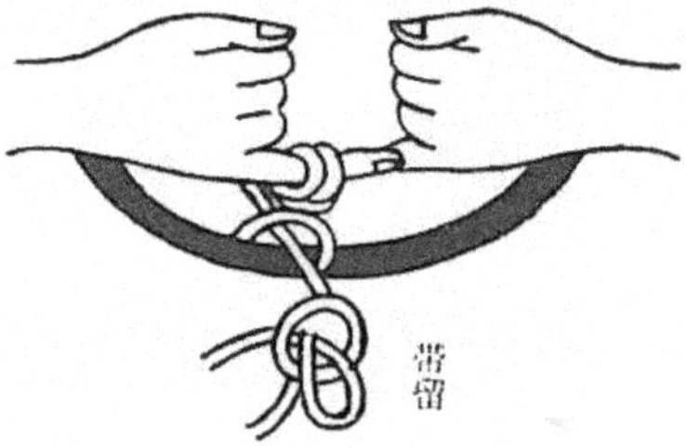

帯ニテ留ル口伝

帯留

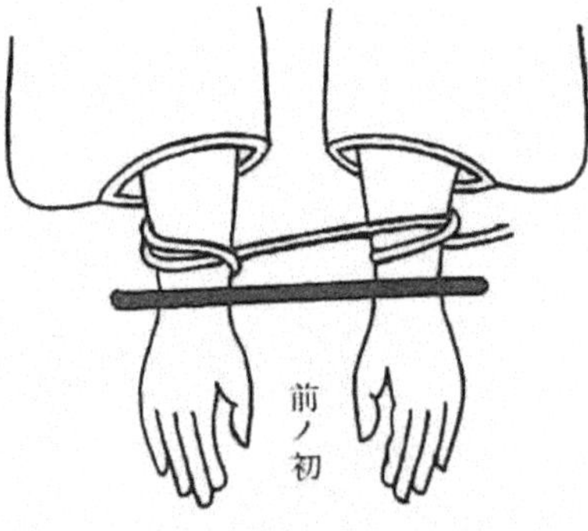

小手附ノ事

前ノ初

Secret Bondage Techniques of the Nanba Ippo & Toh Schools	
八寸縄之事 Hassun Nawa no Koto: How to Use the 24 cm Rope	
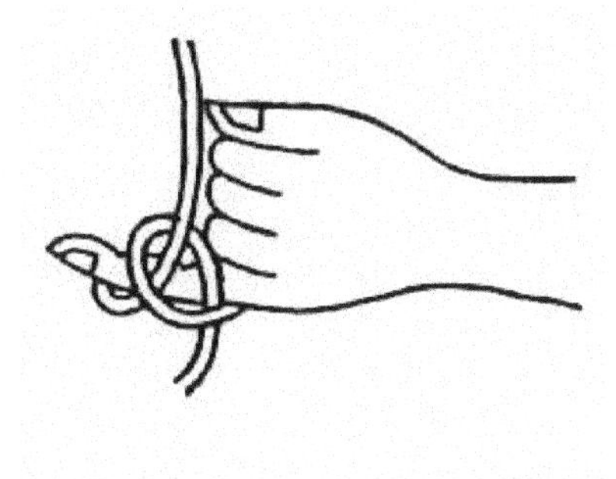	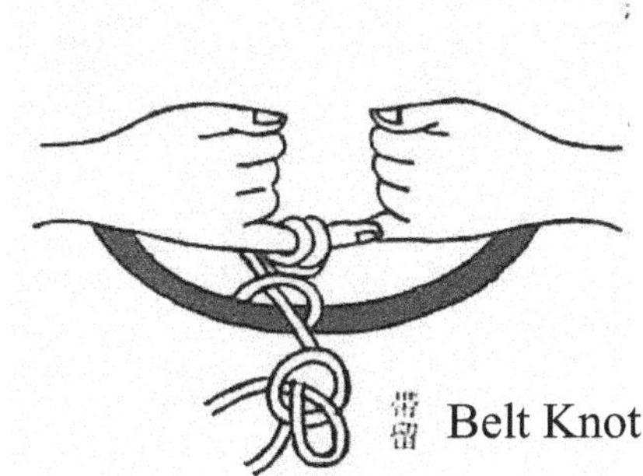
Hassun Nawa no Koto 1: How to Use the 24 cm Rope	Hassun Nawa no Koto 2: 帯ニテ留ルロ伝 There is a Kuden about how to attach the rope to the belt. Note: I presume the black line indicates the belt.
小手付ノ事 Kote Tsuke no Koto: On Restraining The Hands	
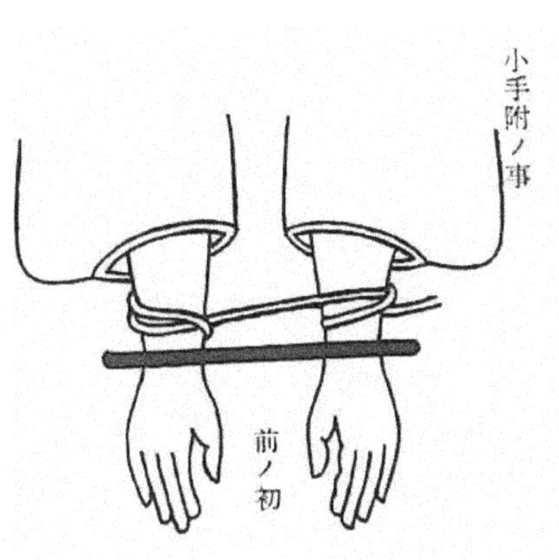 	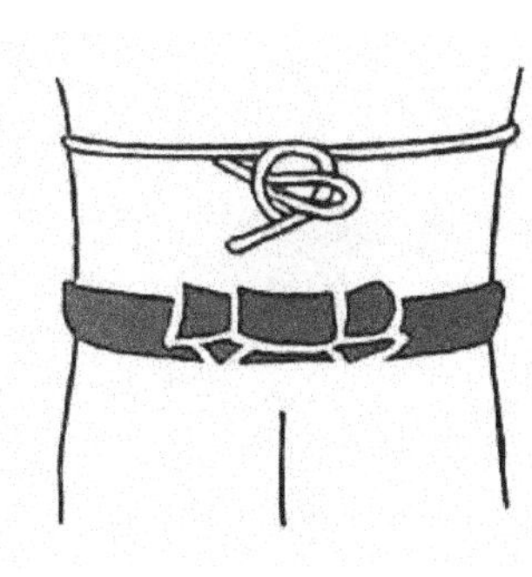
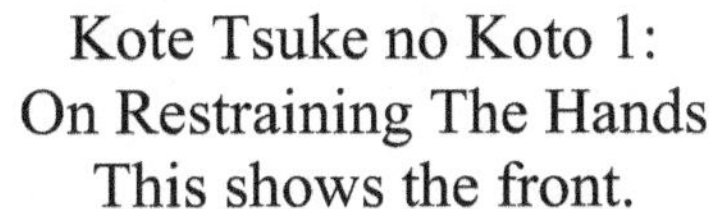 Kote Tsuke no Koto 1: On Restraining The Hands This shows the front.	Kote Tsuke no Koto 2: On Restraining The Hands Back View

Secret Bondage Techniques of the Nanba Ippo & Toh Schools
ハガイ〆ノ事 Hagai-jime no Koto: Double Arm Lock **官僧正縄懸様留様** Kan Zo Sho Nawa Kake Yoh Tome Yoh: Proper Arrest Tie for Abbots of Buddhist Temples **無官出家等縄掛様** Mu Kan Shukke Toh Nawa Kake Yoh: Proper Arrest Tie for Non-Abbot Buddhists and Travelling Monks

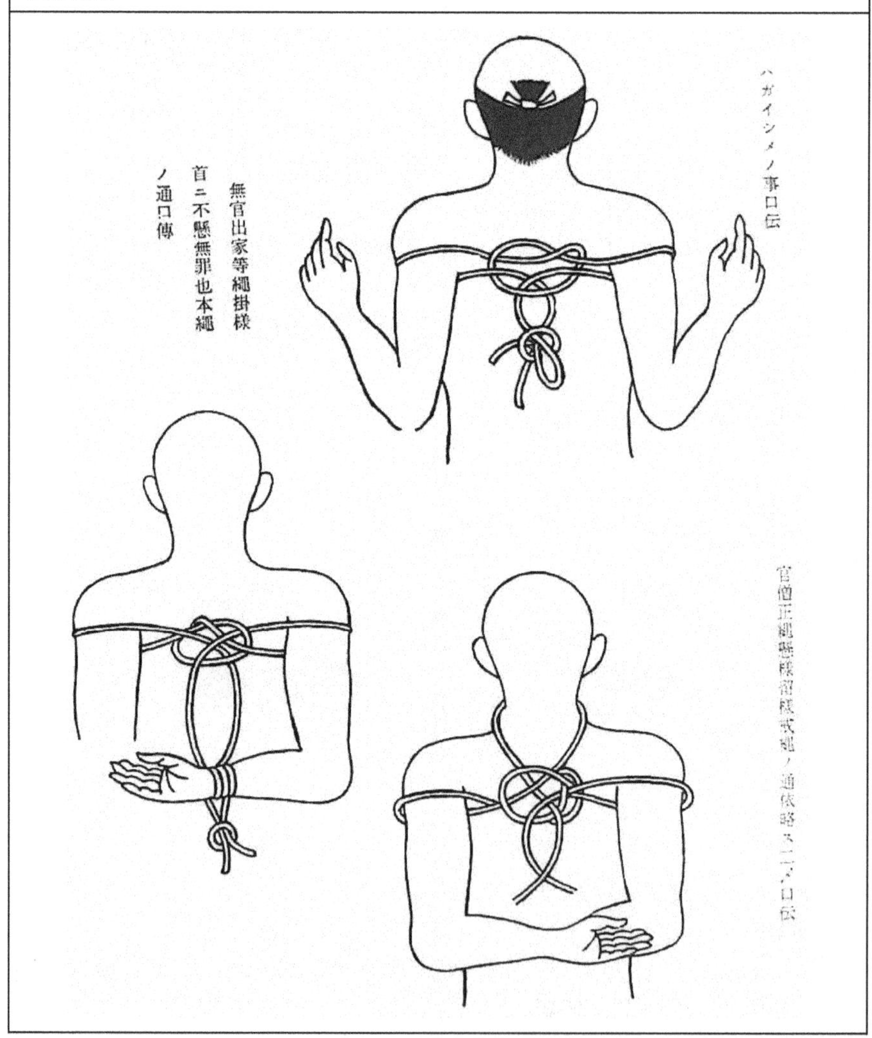

Secret Bondage Techniques of the Nanba Ippo & Toh Schools

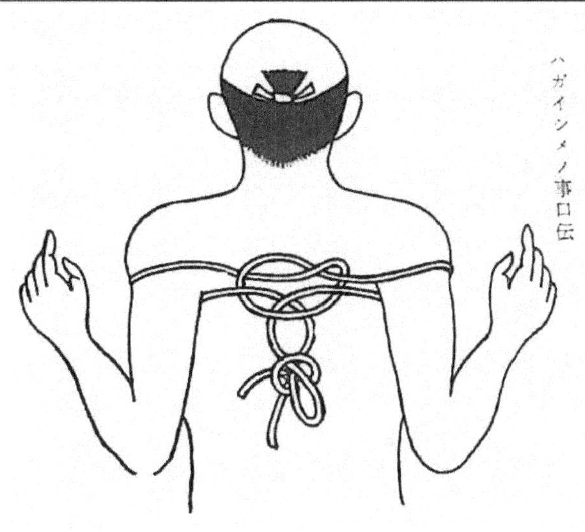

ハガイシメノ事口伝
Hagai-jime no Koto: Double Arm Lock. There is a Kuden.

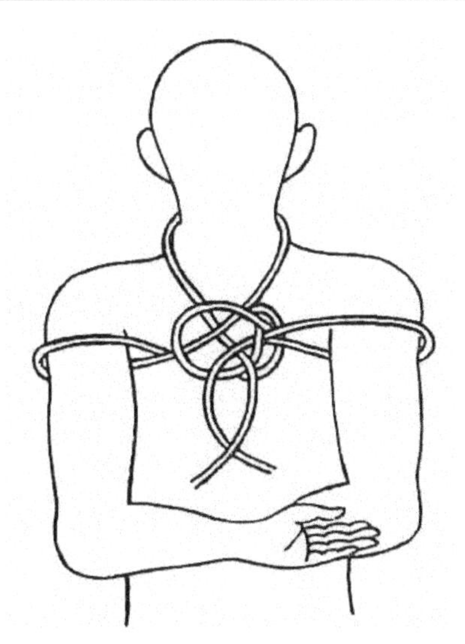

官僧正縄懸様留様
Kan Zo Sho Nawa Kake Yoh Tome Yoh:

Proper Arrest Tie for Abbots of Buddhist Temples

様戒縄ノ通依略スニ〆口伝
This application is the same as Kai Nawa, so only one illustration will be given. There is a Kuden.

Secret Bondage Techniques of the Nanba Ippo & Toh Schools
無官出家等縄掛様 Mu Kan Shukke Toh Nawa Kake Yoh: Proper Arrest Tie for Non-Abbot Buddhists and Travelling Monks **首ニ不懸無罪也本縄ノ通口傳** If there is no rope tied around the neck, then they are innocent. This is the same as Hon Nawa, Main Tie. There is a Kuden.
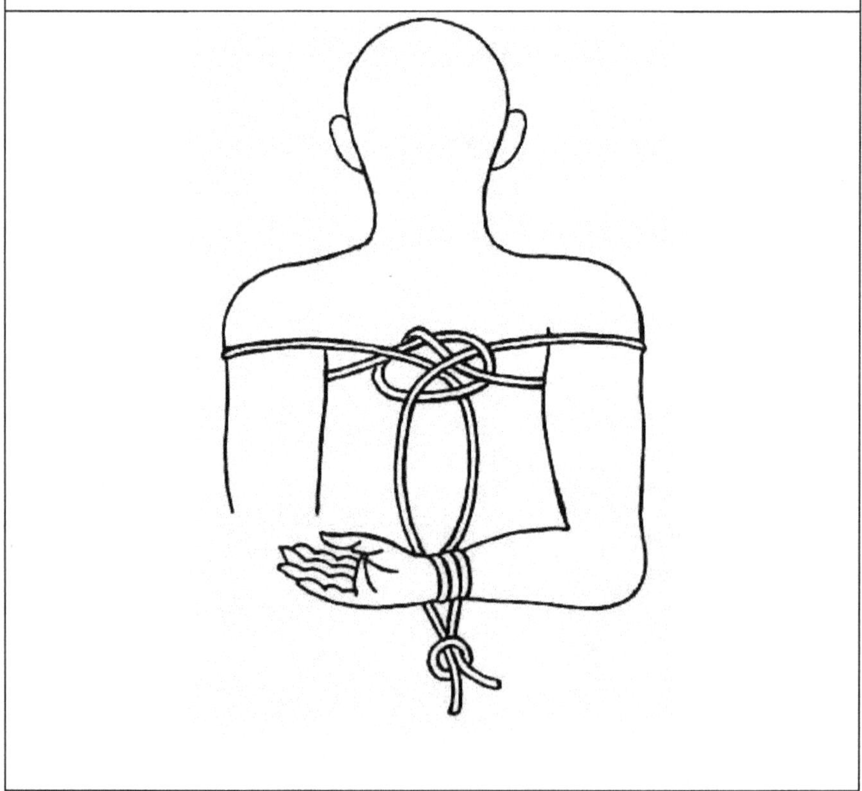

Secret Bondage Techniques of the Nanba Ippo & Toh Schools

死罪ワ縄懸様高手ニ不掛切時縄トクロ傳
留ノ縄追放縄ノ如シ

If a criminal is sentenced for Shizai, a crime carrying a death sentence, then no ties are done on the upper arms. Just before cutting, untie the knot around the neck. There is a Kuden.

This is the same as Tome no Nawa, Stopping Tie and Tsuiho Nawa, Banishment Tie.

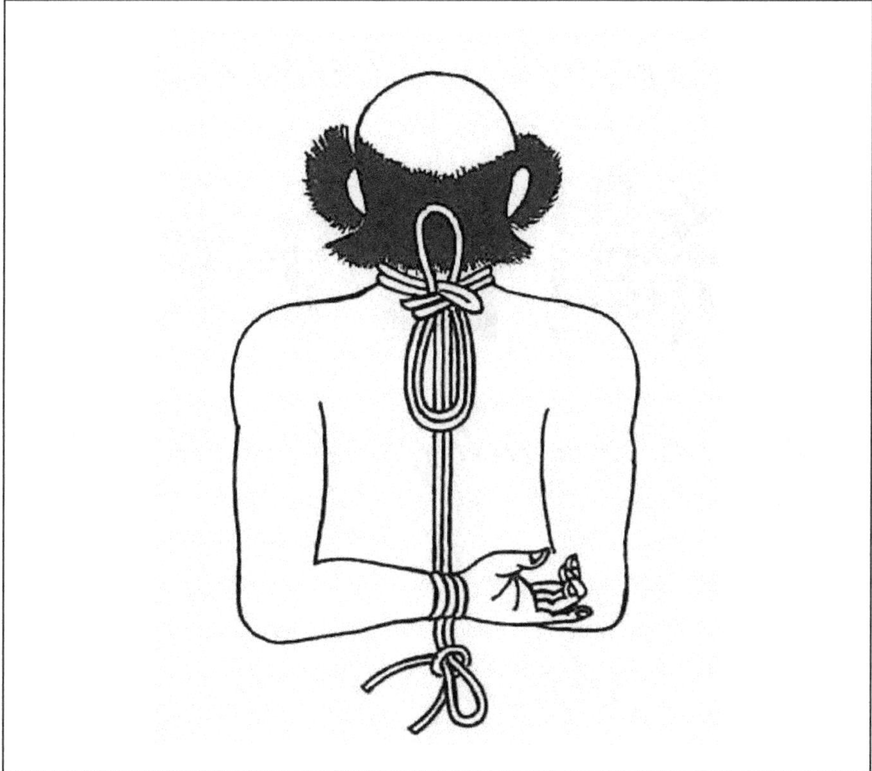

Ichiden Ryu

一伝流

Nawa Mokuroku

縄目録

List of Bondage Techniques of the Ichiden School of Martial Arts

Ichiden Ryu Nawa Mokuroku:
List of Bondage Techniques of the Ichiden School

一傳流縄目録

一 草之本縄不入番
一 行之本縄不入番
一 真之本縄不入番
一 女出家縄
一 神速不思議縄
一 鍵針早縄
一 籠手釣速縄
一 斬縄
一 天狗羽懸縄
一 天狗羽縮縄
右十ヶ条当流表縄也
一伝流極意五ヶ条
一 天結縄　二ヶ条
一 大小下緒縄
一 捕手之事
一 阿弥陀之胸割
一 微塵大極意

Ichiden School	
草之本縄不入番	So no Hon Nawa Ban Irazu: Free Application Main Tie No Guard Needed
行之本縄不入番	Gyo no Hon Nawa Ban Irazu: In Between Main Tie No Guard Needed
真之本縄不入番	Shin no Hon Nawa Ban Irazu: True Application Main Tie No Guard Needed
女出家縄	Onna Shukke Nawa: Travelling Nun Tie
神速不思議縄	Shinsoku Fushigi Nawa: Mysterious God Like Speed Tie
鍵針早縄	Kagi Hari Nawa: Key and Needle Tie
籠手鉤速縄	Kote Kagi Soku Nawa: Fast Wrist Hook Tie
斬縄	Kiri Nawa: Neck Tie/ Headsman's Tie
天狗羽懸縄	Tengu Hanegake Nawa: Mountain Goblin Wing Tie
天狗羽縮縄	Tengu Hane Shuku Nawa: Mountain Goblin Compression Wing Tie
右十ヶ条当流表純也 The previous 10 techniques comprise the Omote, or Outer Ties, of the Ichi Den School	
一伝流極意五ヶ条 The following 5 are the Gokui, the Secret Inner Mystery techniques, of Ichi Den School.	
天結縄二ヶ条	Amaketsu Nawa Ni ka Jo: Heavenly Knot Tie, Two Techniques
大小下緒純	Daisho Sage-O Nawa: Using the Cords on the Long and Short Sword to Tie
捕手之事	Torite no Koto: Arresting Techniques
阿弥陀之胸割	Amida no Mune Wari: Split Chest Amida Buddha Tie
徴僅大極意	Mijin Dai Gokui: Great Inner Mystery Pulverizer Tie

Ichiden School

草之本縄不入番 So no Hon Nawa Ban Irazu: Free Application Main Tie No Guard Needed

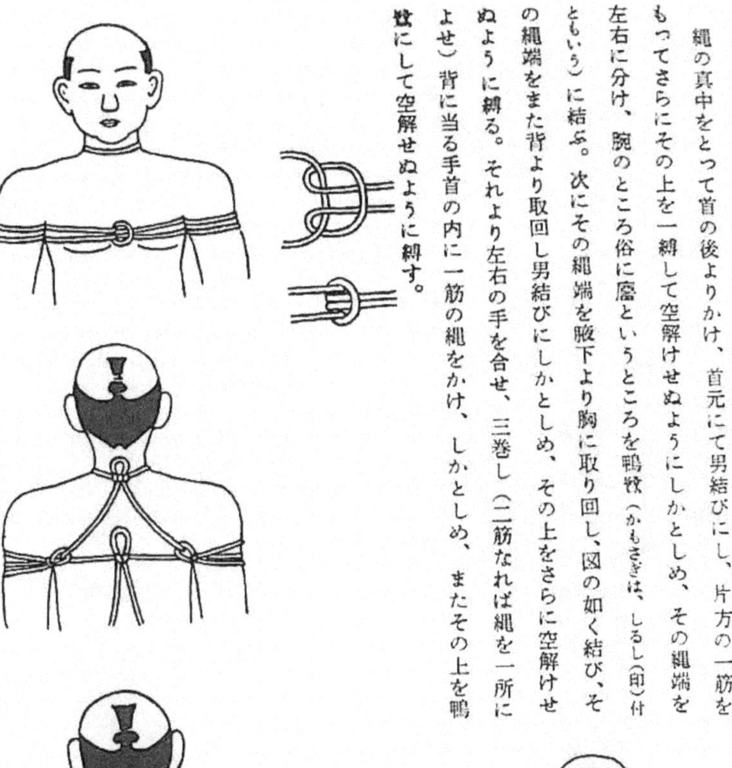

草之本縄不入番

縄の真中をとって首の後よりかけ、首元にて男結びにし、片方の一筋をもってさらにその上を一締して空解けせぬようにしかとしめ、その縄端を左右に分け、腕のところ俗に腐というところを鴨敓（かもさぎは、しるし（印）付ともいう）に結ぶ。次にその縄端を腋下より胸に取り回し、図の如く結び、その縄端をまた背より取回し男結びにしかとしめ、その上をさらに空解けせぬように締る。それより左右の手を合せ、三巻し（二筋なれば縄を一所によせ）背に当る手首の内に一筋の縄をかけ、しかとしめ、またその上を鴨敓にして空解せぬように締す。

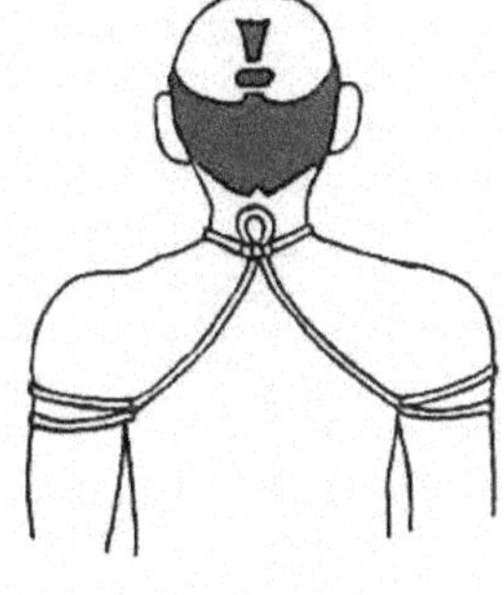

Ichiden School
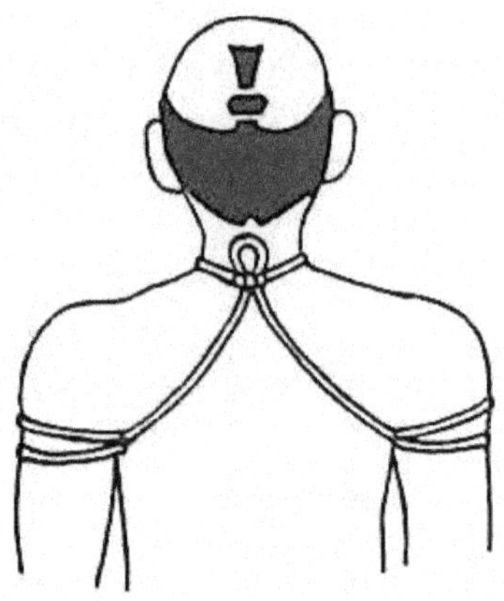
Fold the rope in the center and wrap it around the prisoner's neck from behind. At the base of his neck tie a Men's Knot. In addition, using one strand of rope, tie a loop on top of the knot on his neck. Split the rope into 2 strands and tie a Kamo-Sagi, Duck-Heron Knot, on the arms. The spot the rope is tied on is known colloquially as [illegible Kanji] The Duck-Heron Knot is also known as Tsurushi Tsuke, Making a Mark. After that bring one strand under each armpit and tie the knot shown in the illustration below.
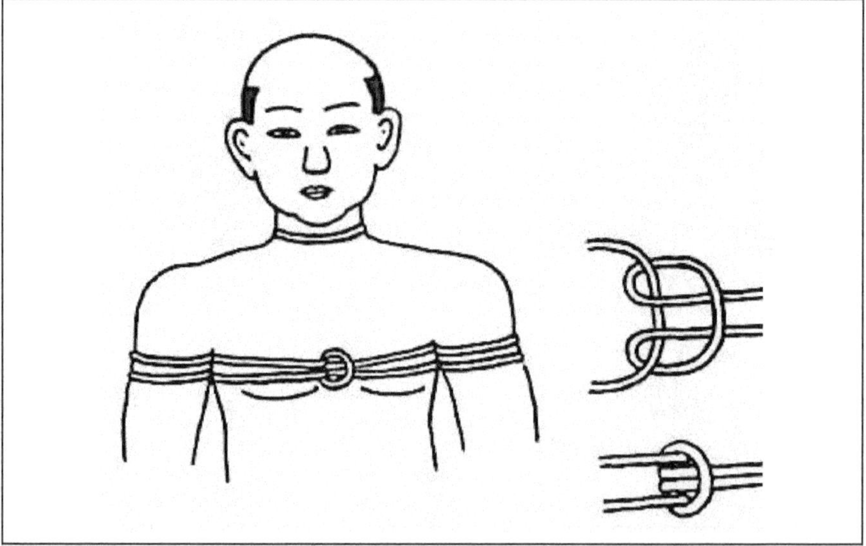

Ichiden School

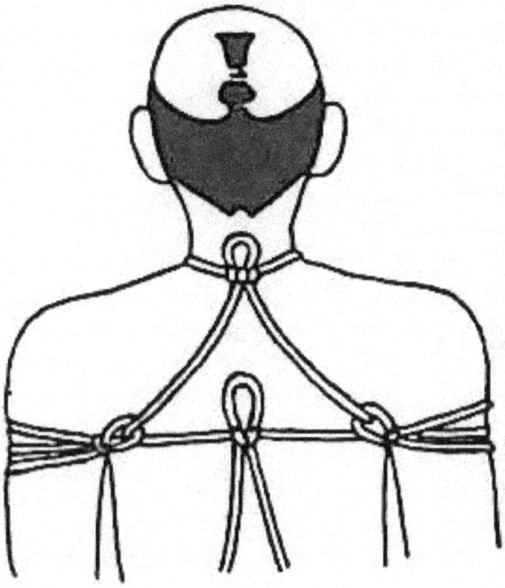

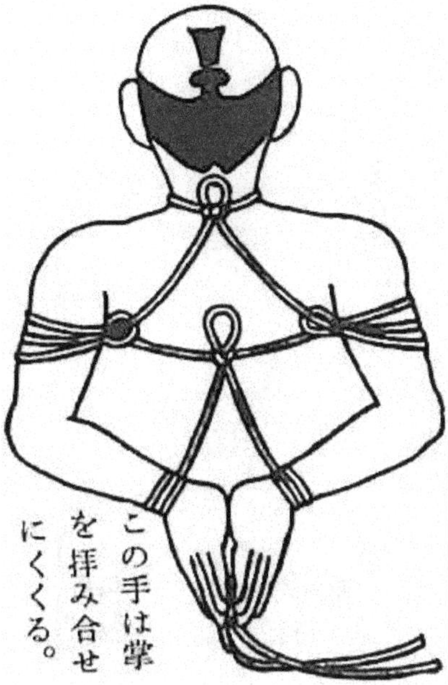

Pass the two strands around the back and tie in a Men's Knot in the center of the back. On top of that knot tie a loop.
Next, bring the left and right hands together and wrap the wrists three times. If you only have a single strand and don't have a doubled rope, then overlap the wrists and tie in one place. With the wrists against the back take one strand and secure the wrists to the back (not shown?) then tie a Duck-Heron Knot and a loop on top of that.
Press the prisoner's palms together so it looks he is praying. Finally, wrap the rope around the hands.

Ichiden School

真之本縄不入番草　Shin no Hon Nawa Ban Irazu:

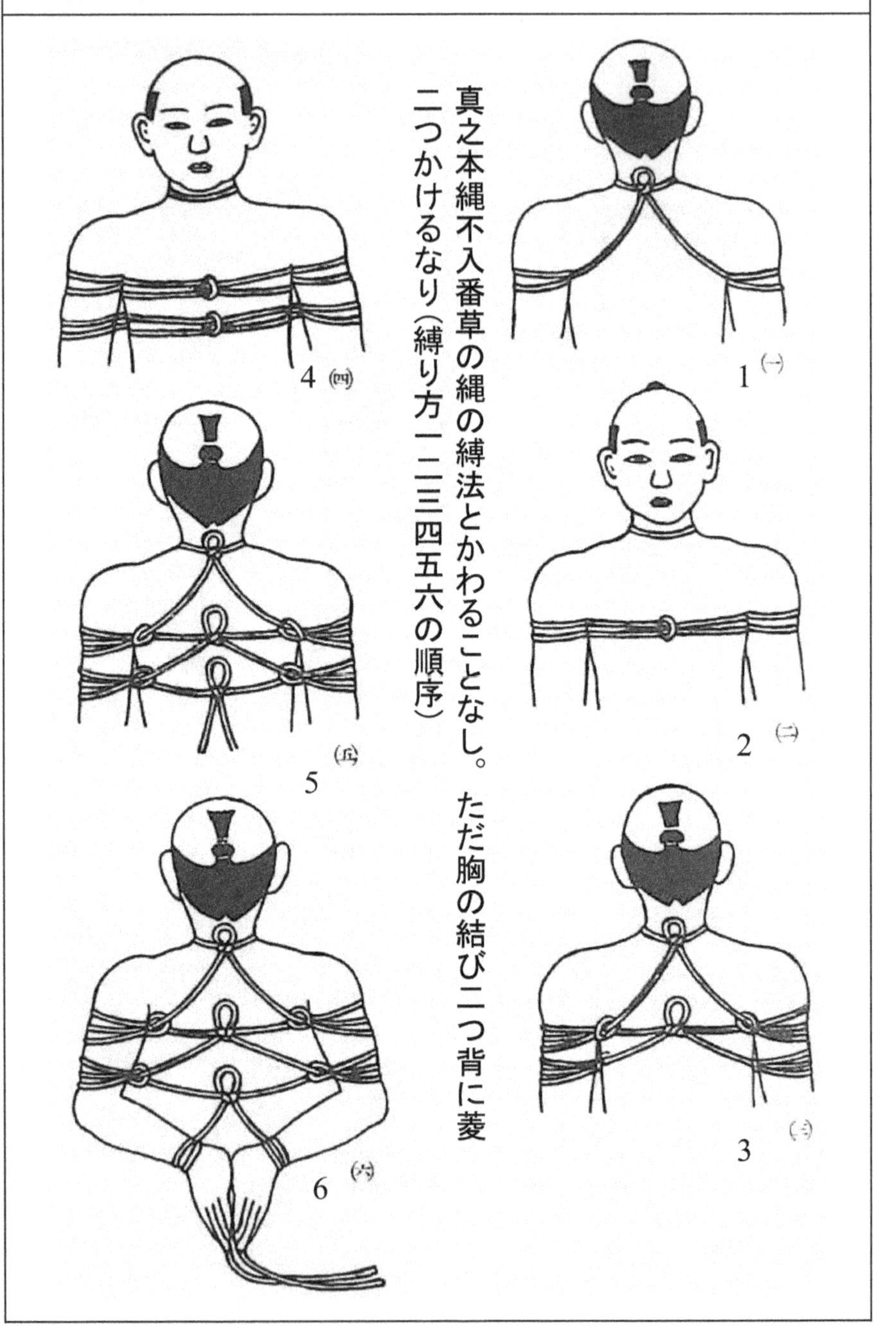

Ichiden School

真之本縄不入番草　Shin no Hon Nawa Ban Irazu:
How to Tie True Application Main Tie No Guard Needed.

真之本縄不入番草の縄の縛法とかわることなし。ただ胸の結び二つ背に菱二つかけるなり（縛り方一二三四五六の順序）

This method is the same as the previous So no Hon Nawa Ban Irazu: Free Application Main Tie No Guard Needed, except that there are two ties on the chest and two caltrop ties on the back. The order of how to tie goes 1, 2, 3, 4, 5, 6.

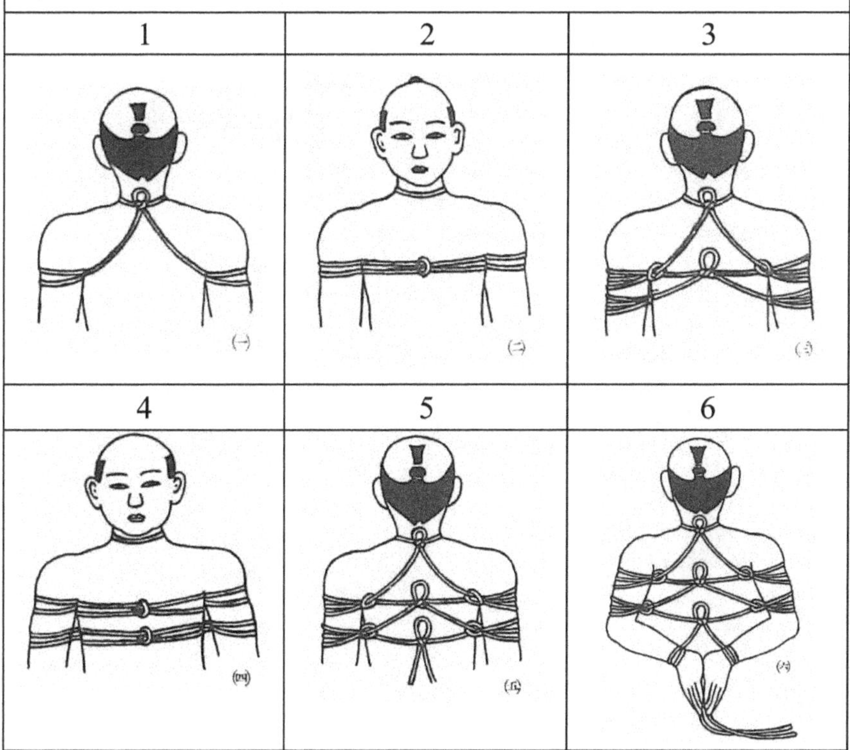

Ichiden School

女出家縄 Onna Shukke Nawa: Female Travelling Monk Tie
当流ではこの縄を以て女人と出家沙門を兼縛る。
This is the tie used in Ichiden Ryu to restrain females who are also travelling nuns.

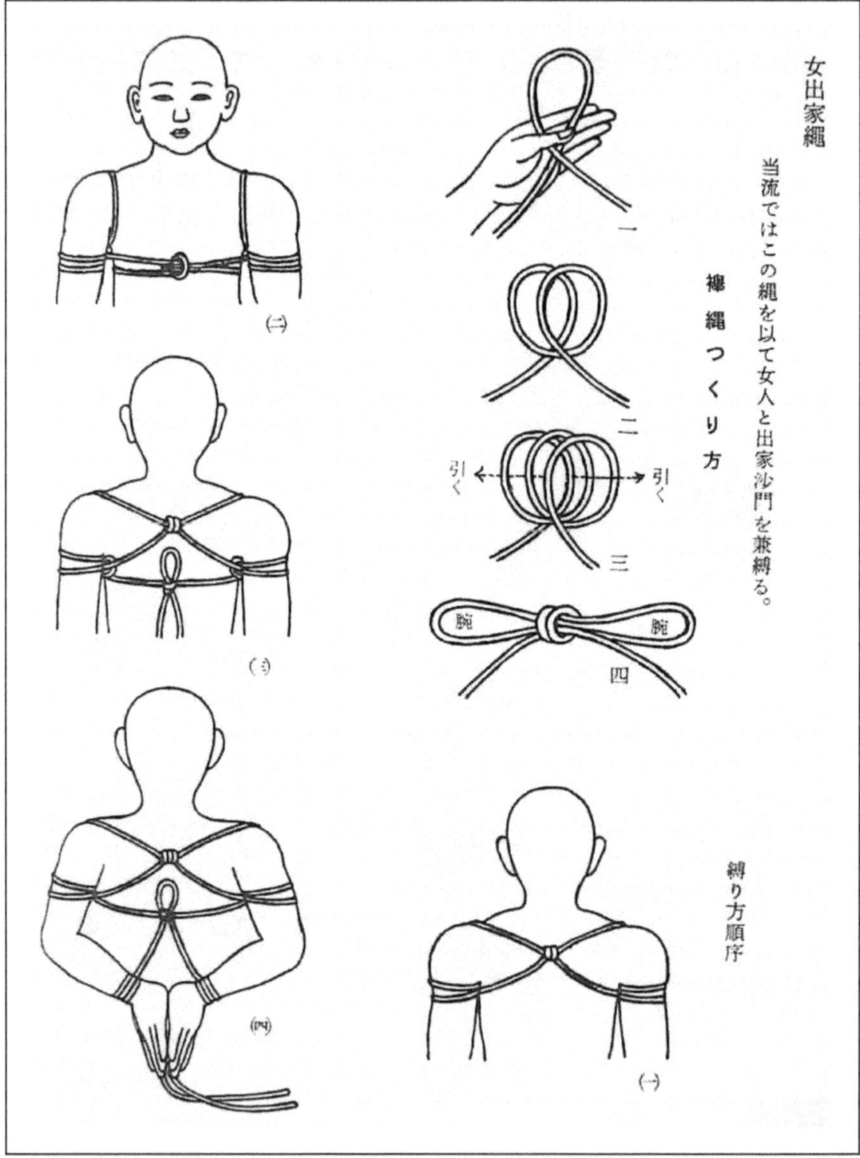

Ichiden School

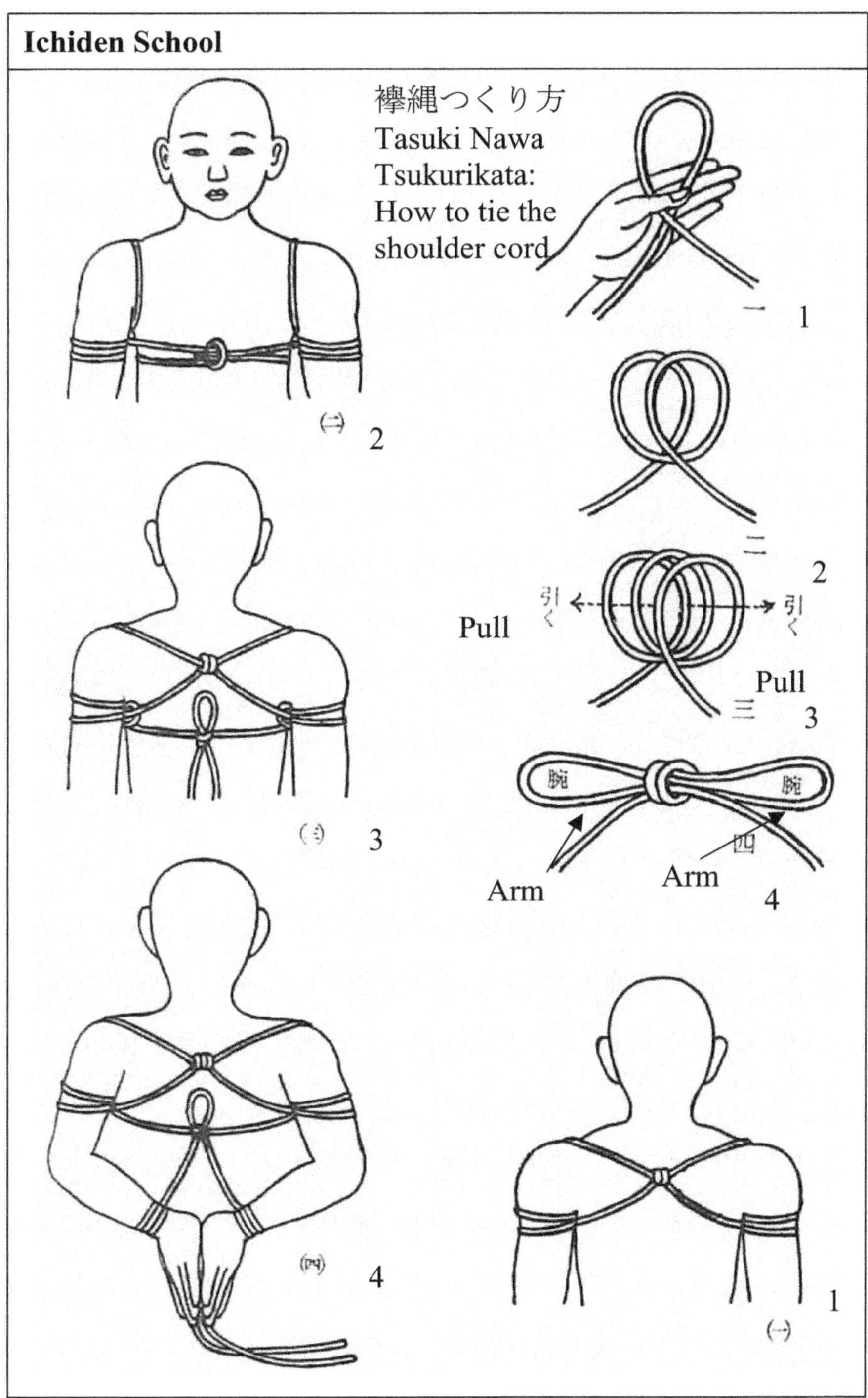

Notes:
There are several different definitions for Onna Shukke.

1. A woman who commits herself to the life of a nun and follows the Guzoku Kai 具足戒(this is another word for the Jukai previously mentioned,) the rules and regulations of a Bikuni or nun There are 348 rules for women (and 250 for men.)

2. During the middle ages of Japan (12th ~16th centuries) the word described entertainers who dressed as nuns and travelled around Japan performing.

3. During the Edo Era 1600-1868 the word referred to low ranked courtesans who dressed as nuns.

Originally women were forbidden to work as travelling nuns however, eventually they were permitted. Rules and regulations were written for Bikuni 比丘尼 and dedicated temples founded. Initially, girls from wealthy families made up the majority and the places they resided became known Bikuni Gosho 比丘尼御所, or nunneries.

The rules for women were similar to those for men. Women had to do Teihatsu Nyudo 剃髪入道, Shave the Head, Enter the Monastic Life, and follow other strict rules. However, as time went on the rules became relaxed and some girls remained at home and only cut their hair short, instead of shaving their heads. They spent their days proselytizing using illustrations of heaven and hell and singing spiritual songs. Eventually, these Uta-Bikuni 歌比丘尼 singing nuns became a sort of poor underclass in the Edo Era. The singing nuns evolved into the low ranked courtesans that worked the streets calling out to passersby (as opposed to working in the official red-light district of Yoshiwara.) The Uta-Bikuni also operated small boats on the side of the river and would invite clients onto their boat.

Top: Illustration of a Hikuni nun
Bottom: Illustration of a Uta Bikuni, singing nun, calling out to a potential client.
Illustrations from *World Encyclopedia* 世界大百科事典 1968.

Ichiden School
神速不思議縄 Shinsoku Fushigi Nawa: Mysterious God Like Speed Tie Note: The title written here is slightly different from the initial list. **神変不思議縄** Shinpen Fushigi Nawa: Mysterious God Like Multitude of Changes Tie. Note: It is not clear if this is an alternate name.

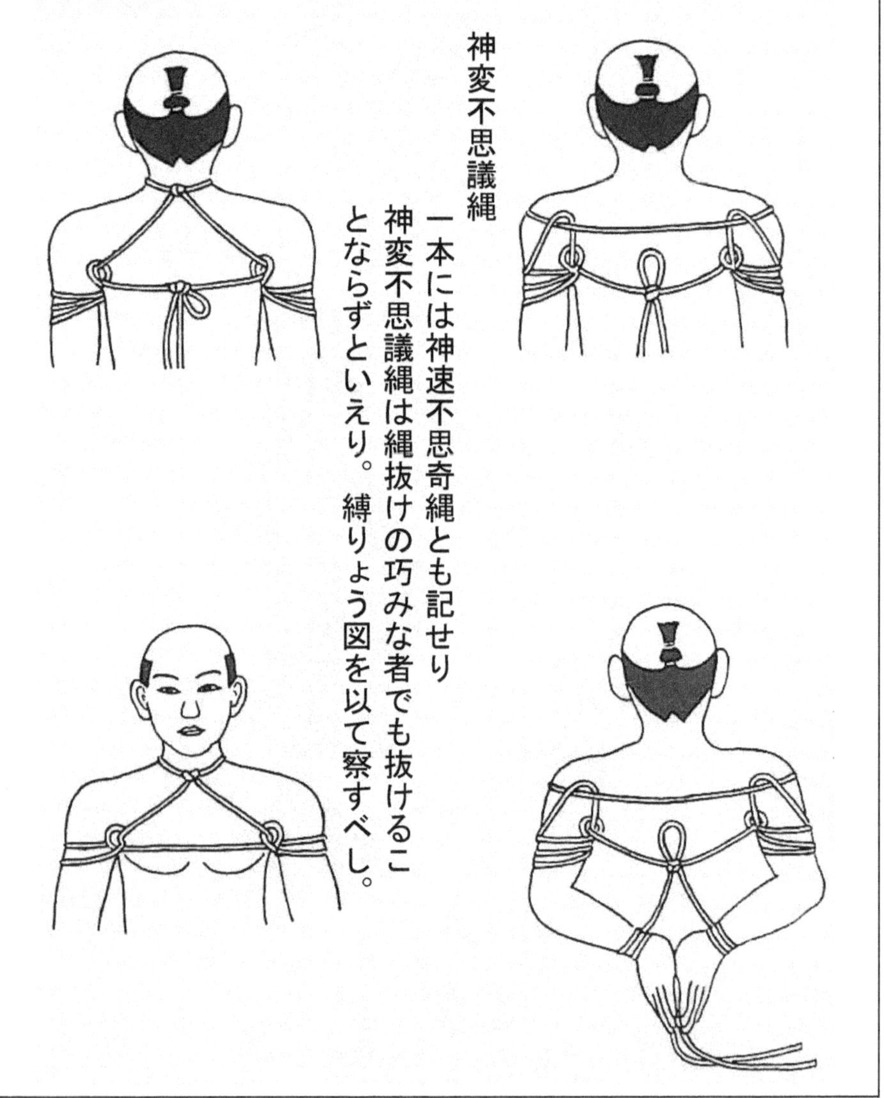

Ichiden School

神速不思議縄
Shinsoku Fushigi Nawa: Mysterious God Like Speed Tie

This mysterious tie goes on in a flash of divine speed. Even those adept at escaping ties will find they cannot free themselves. Consult the four illustrations below for how to tie.

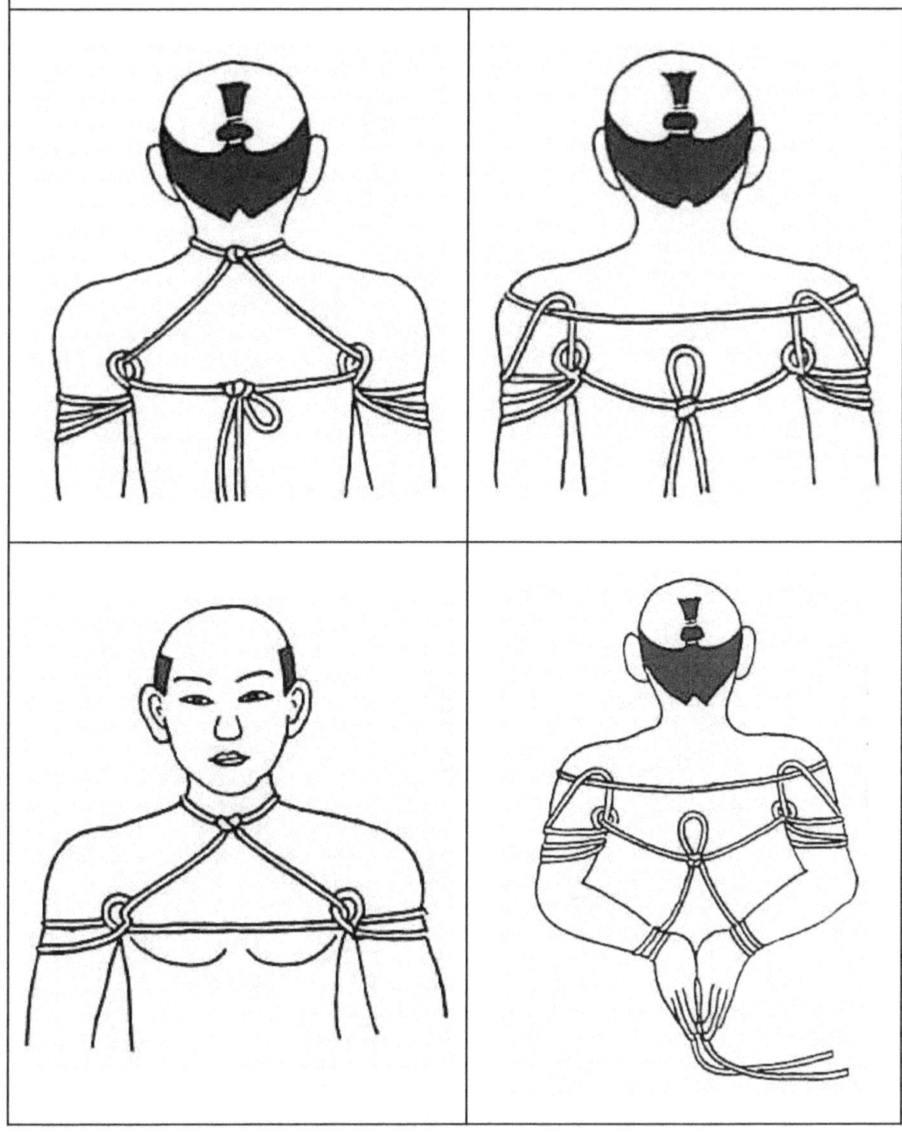

Ichiden Ryu Hiden: Secret Teachings of the Ichiden School

Kagi Hari: Key (Hook) Needle

The measurements for this are Kuden.
When doing Hon Nawa, Main Tie, the rope is 7 1/2 arm-spans long.
When doing Haya Nawa, Fast Tie, the rope is 1 Jo 7 Shaku.
There is also a Haya Nawa, Fast Tie, with no hook.
There are Kuden for each of the above.

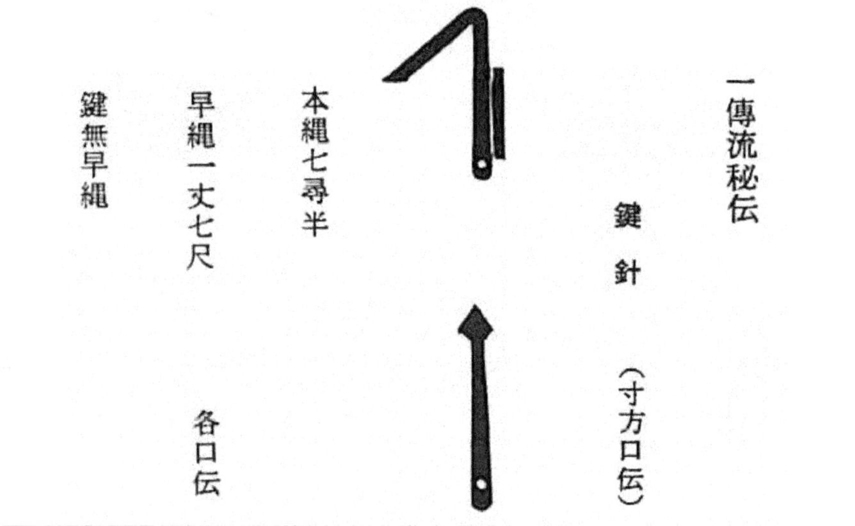

Kagi Hari Haya Nawa: Key Needle Fast Tie

There are a variety of ways to hook the person.

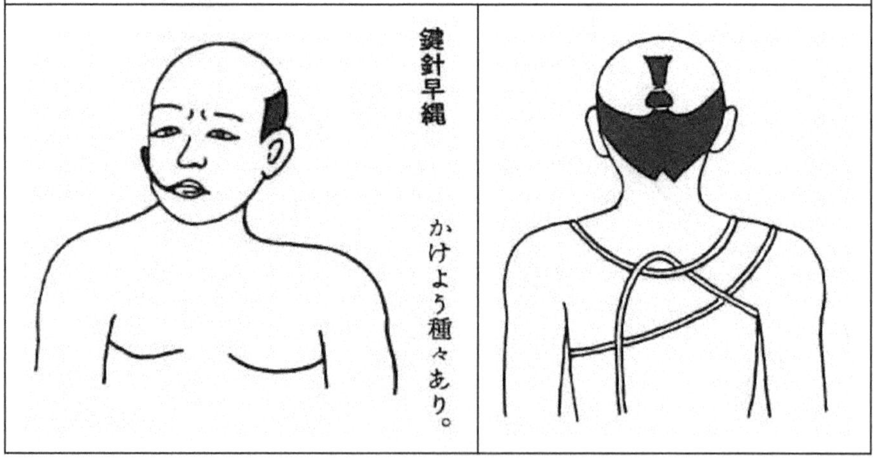

Ichiden School

籠手釣早縄

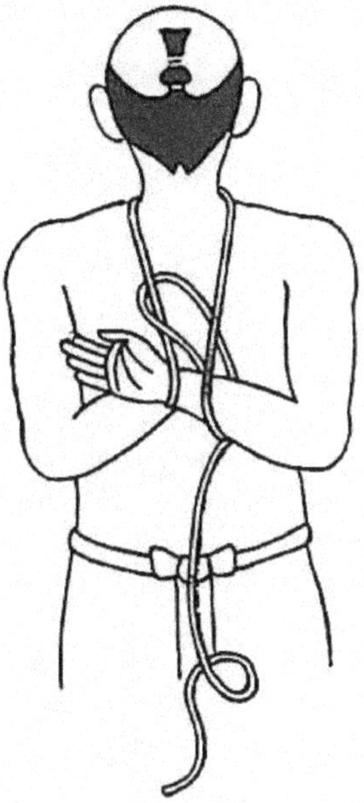

縄を腕首に引かけ、その縄を首に回し、片方の手をとって先の手に重ねてしめる。

斬縄

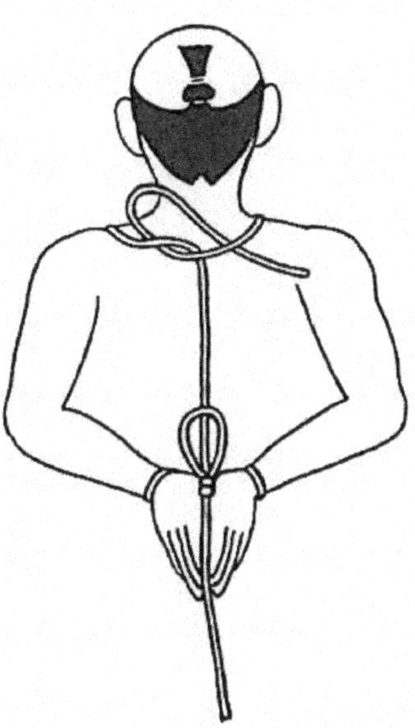

この縄は打首の者にかける縄にて、打首の時、首の縄端を引けばすぐ解ける。縄を解くところは手計りなり。

Ichiden School	
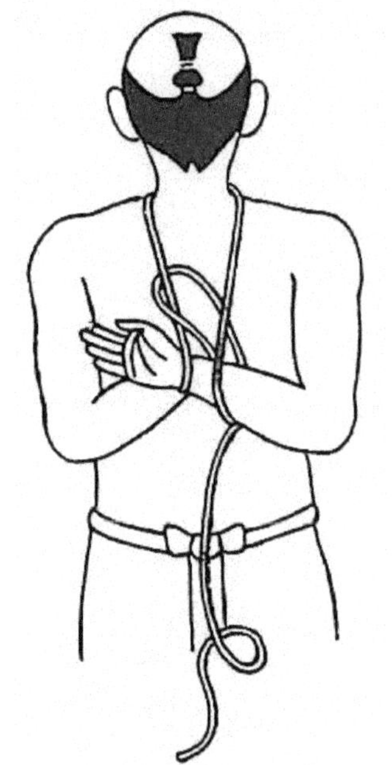	籠手鉤速縄 **Kote Kagi Soku Nawa:** **Fast Wrist Hook Tie** Wrap the rope around the wrists and then once around the neck. Take one hand and place it on top of the other before tying them together.
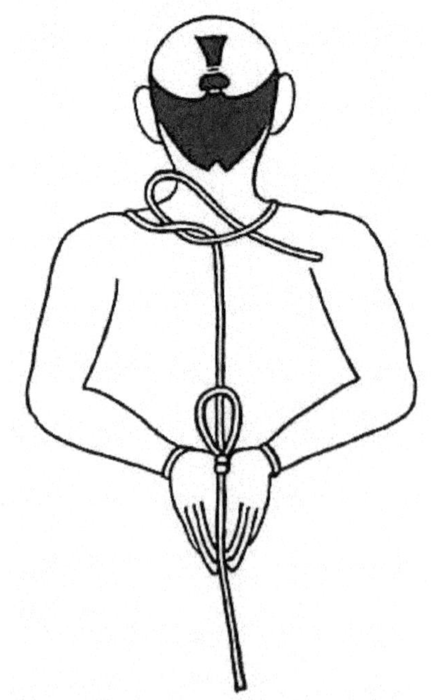	斬縄 **Kiri Nawa:** **Cut Tie/ Headsman's Tie** This tie is done for prisoners sentenced to Uchi Kubi, beheading. When the execution is set to begin, the rope around the neck should be quickly released. It is important that this rope be released deftly.

Ichiden School

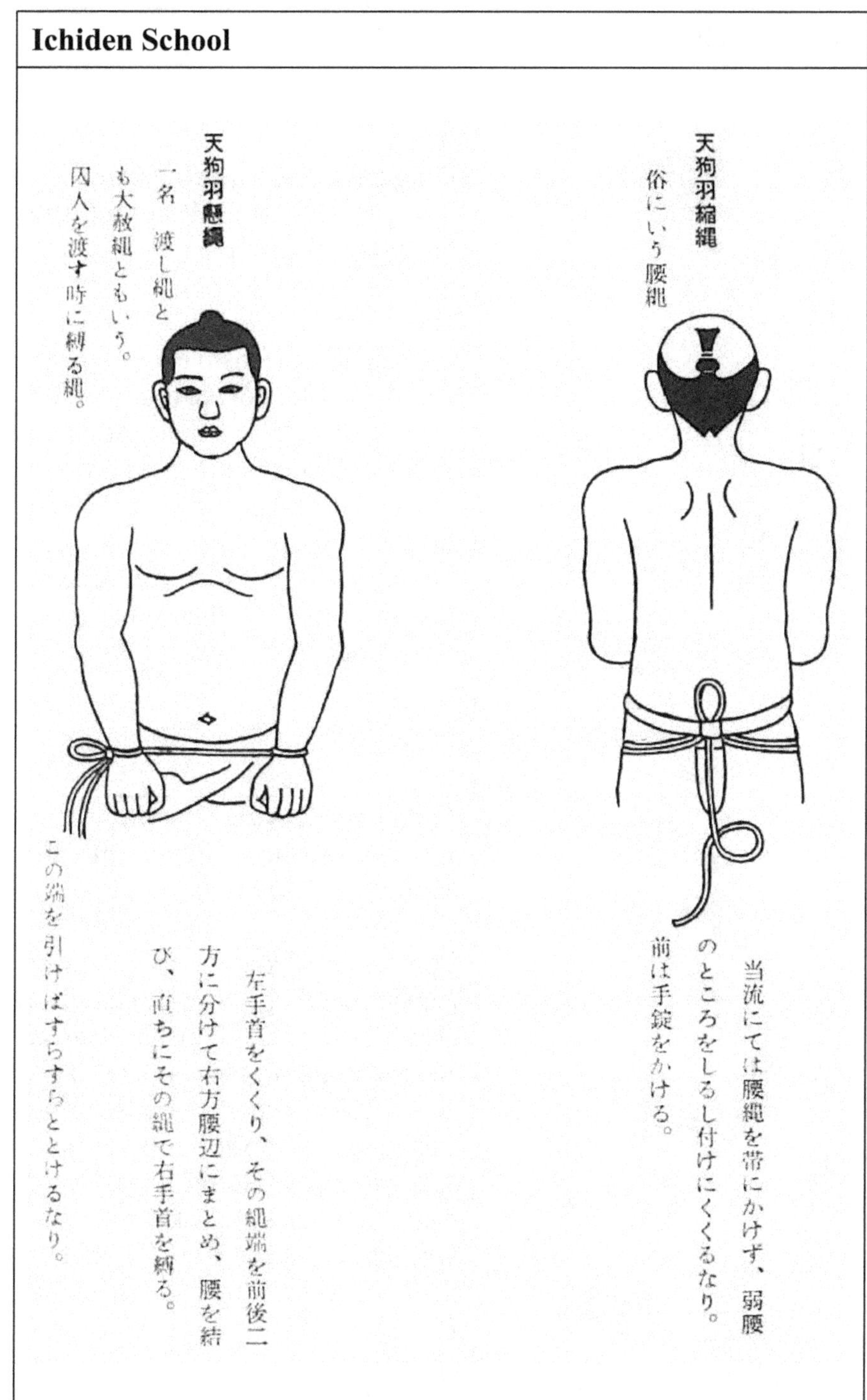

Ichiden School	
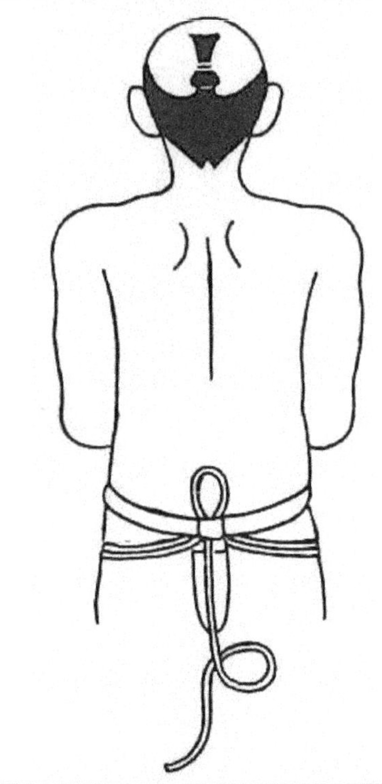	**Tengu Hane Chijimu Nawa: Compressing the Wings of the Mountain Goblin** This technique is known colloquially as Koshi Nawa, Waist Tie. In this school we don't tie the rope to the Obi, or belt, rather to Yowa Goshi, the point where the Obi is loosest, at the small of the back. A loop is passed through Yowa Goshi and then the rope is passed around to the front to tie rope handcuffs.
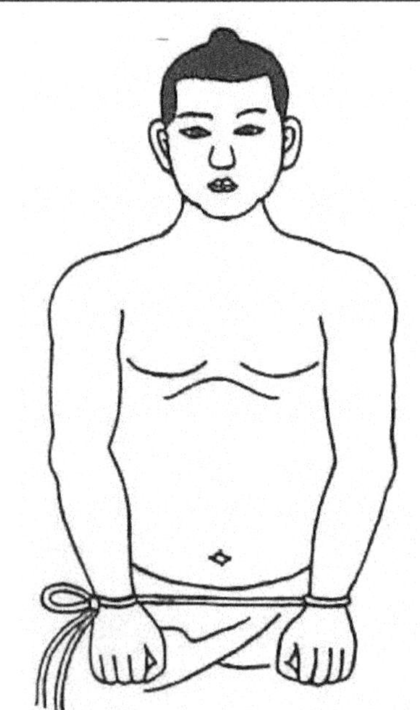	**Tengu Hane Kake Nawa: Restraining the Wings of the Mountain Goblin** This is also known as Watashi Nawa, Handing Over Rope. It is also known as Taisha Nawa, Beseeching Forgiveness Tie. This is used when transferring prisoners. Wrap the rope around the left wrist. Split the strands into front and back. Take one end around his back and tie a knot at his waist. Then tie the left wrist. If you pull this strand it will come apart easily

Ichiden School

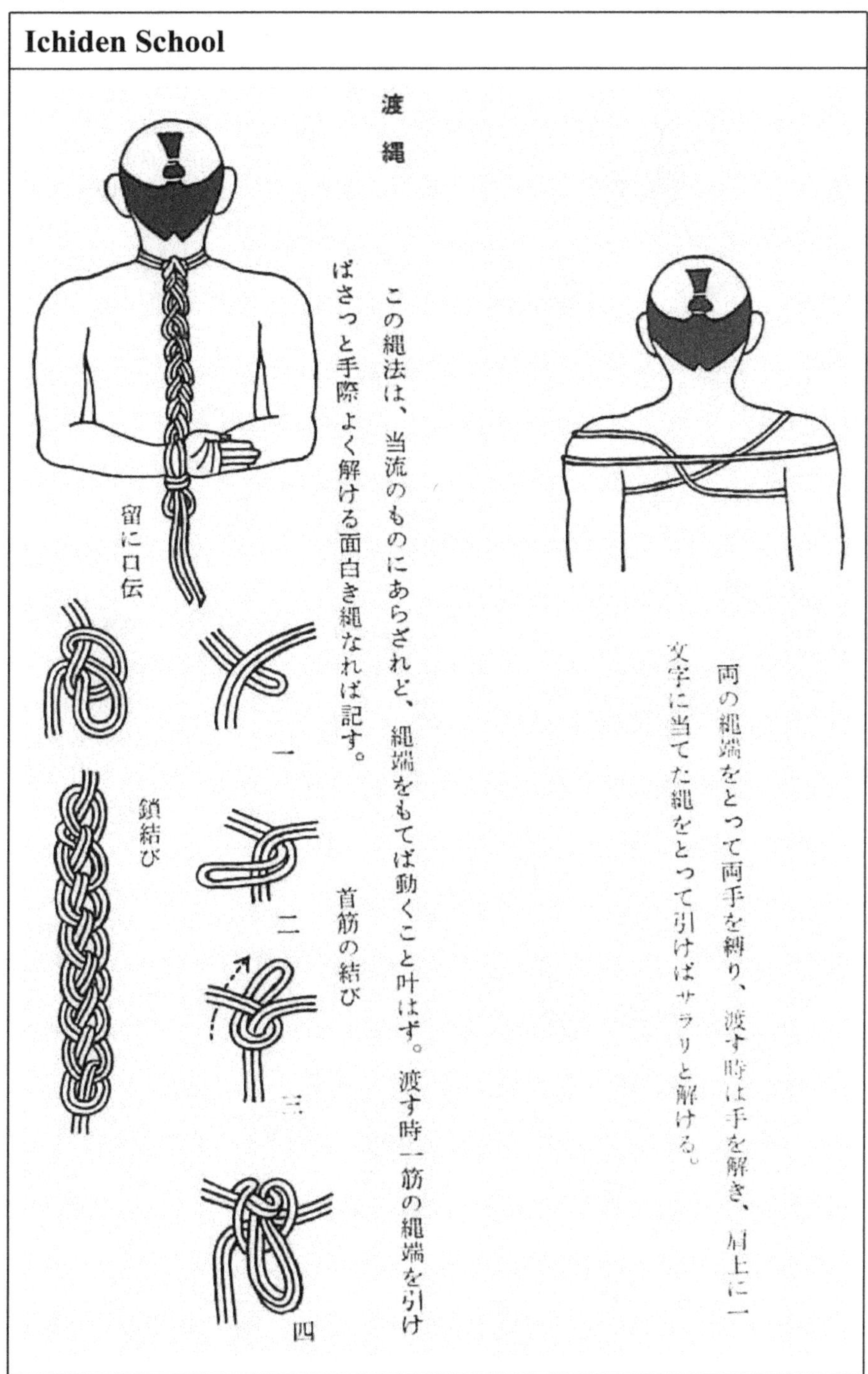

Ichiden School	
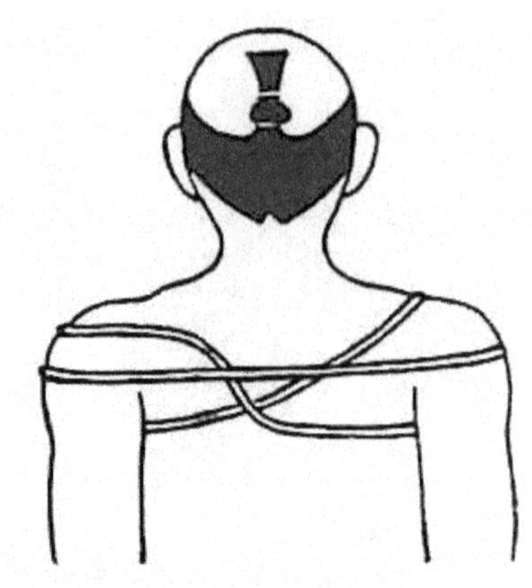	**No Title?** Use the ends of the rope to tie both hands. When handing the prisoner over, release the bonds. If you take hold of the piece of rope making an Ichi Monji, or a straight line like the Kanji for one 一, then the tie will fall smoothly apart.
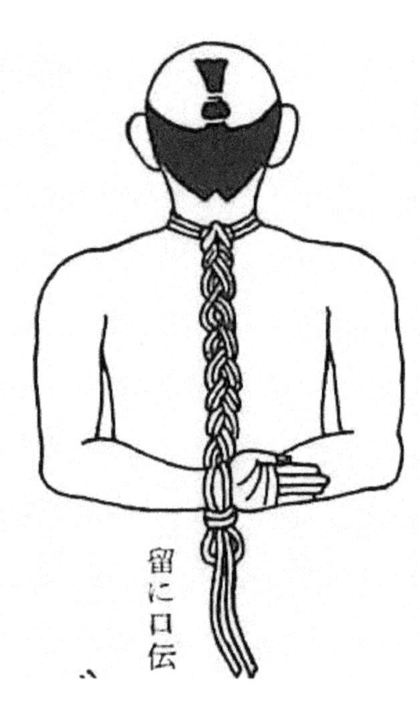	**Watashi Nawa** **Handing Over Tie** This Nawa Ho, Tying Method, is not a technique from this school. If you are holding on to the end of the rope after tying up the prisoner, they will be rendered unable to move. All you need to do when handing the prisoner over is pull on a single strand. This will result in the knot falling apart smoothly in a thoroughly satisfying manner. For that reason, it has been included in this scroll. How to tie the first knot at the neck. Chain Tie The final tie is Kuden.

Ichiden School

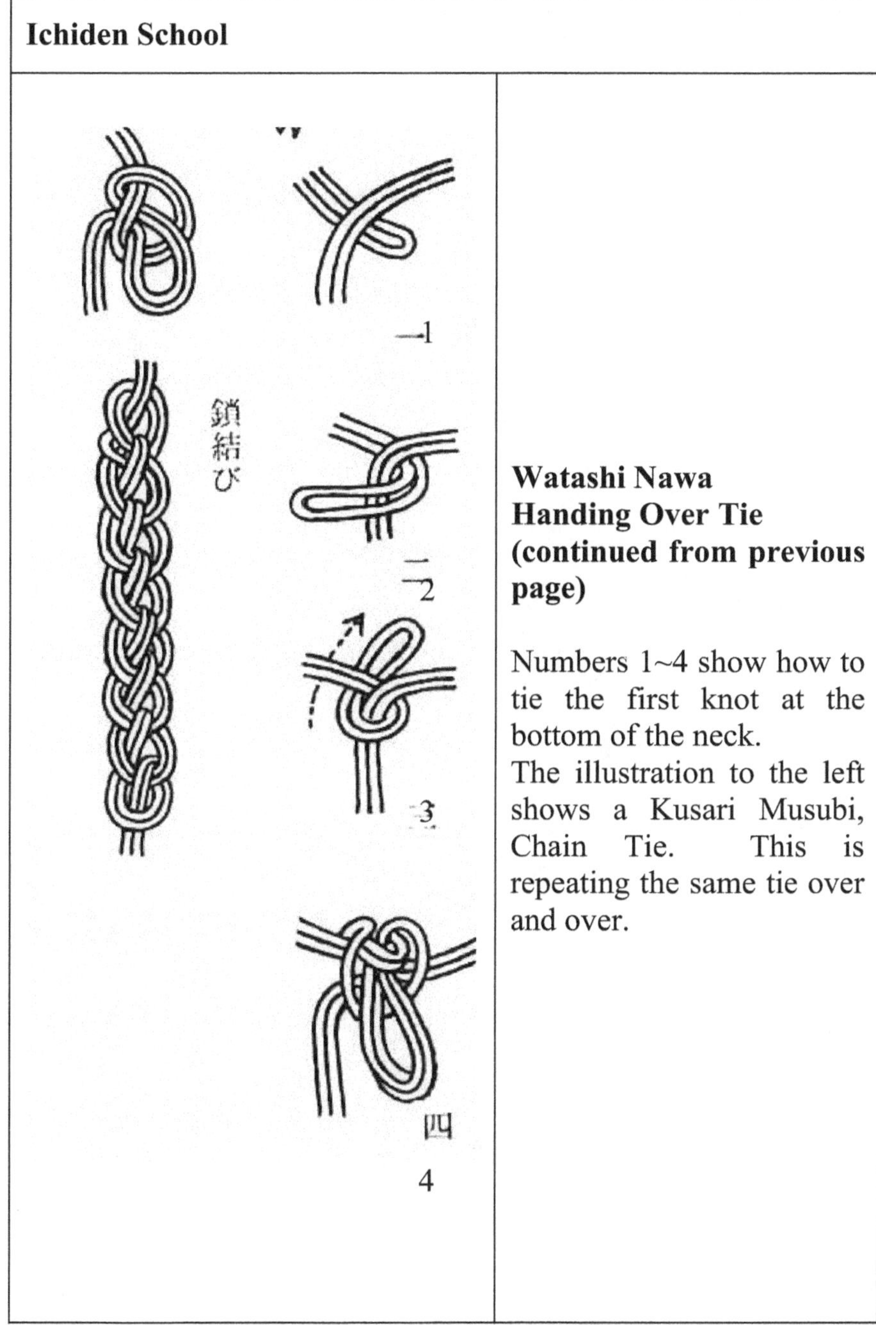

Watashi Nawa
Handing Over Tie
(continued from previous page)

Numbers 1~4 show how to tie the first knot at the bottom of the neck.

The illustration to the left shows a Kusari Musubi, Chain Tie. This is repeating the same tie over and over.

Ichiden School

Moto Yui Nawa: Tie of Twisted Paper

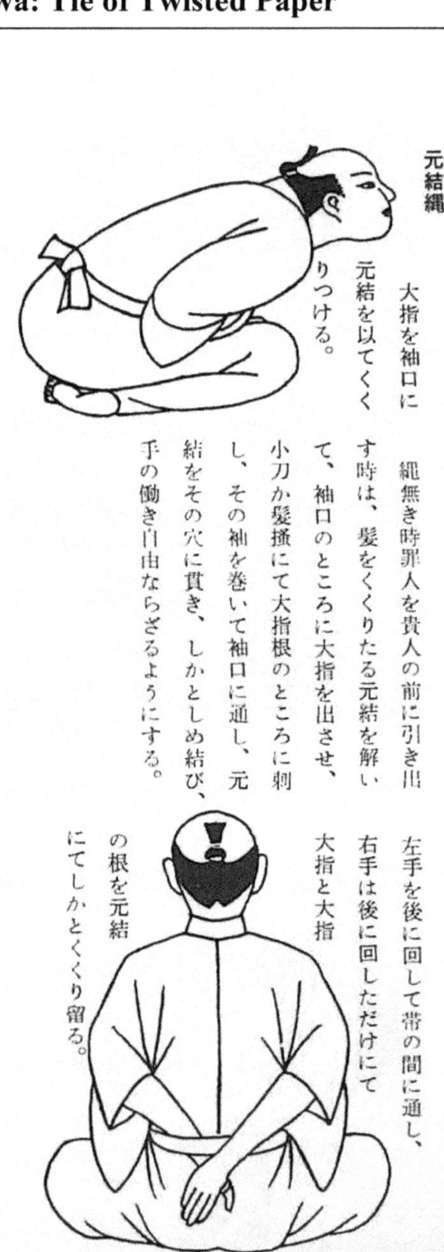

Ichiden School
Moto Yui Nawa: Tie of Twisted Paper #1

This technique wraps the thumbs in the sleeves and secures them with a piece of Moto-yui, paper cord used to tie up the Samurai topknot. This technique is used when you don't have any rope and will have to move a criminal in front of nobles. Untie the Moto-yui tied around [your? his?] hair and pull his thumbs out of his sleeves. Note: I think this means have the prisoner pull his arms inside his Kimono top and then just have his thumbs extend out. Slide the short sword across the palms where they meet the thumb. If a short sword is not available, a Kami Kaki, hairpin/ head scratcher can be substituted. Wrap the sleeves around the hands. Slide the Moto-yui cord through the opening in the sleeve and pass it all the way through to the other side and tie it off securely. Ensure that the hands can't move.

Ichiden School
Moto Yui Nawa: Tie of Twisted Paper #2

Another version is to bring his left hand behind him and pass it through his belt. Bring his right hand behind him and join his thumbs together. Use the Moto-yui cord to tie the bases of the thumbs together.

Ichiden School

Daisho Sageo Shibari
Using the Cord Attached to Either the Long or Short Sword

大小下緒縛り

Ichiden School

Daisho Sageo Shibari
Using the Cord Attached to Either the Long or Short Sword

罪人動く時は柄頭を背に突き当てるべし。

縄なき時、大小の下緒を以て縛る術なり。
強力の者はこのようにて下緒の水玉緒の内に両腕を差し入れさせ、下緒をねじりて、柄頭にて背筋を突くようにすると動くこと叶うべからず。
また手を搦上げて、髪の結び目に脇差の反角をかけ置く時はまた動き得ず。

鞘を引き上げれば手上がる。その時髪に反角をかけ置く。

Ichiden School
Daisho Sageo Shibari #1 Using the Cord Attached to Either the Long or Short Sword
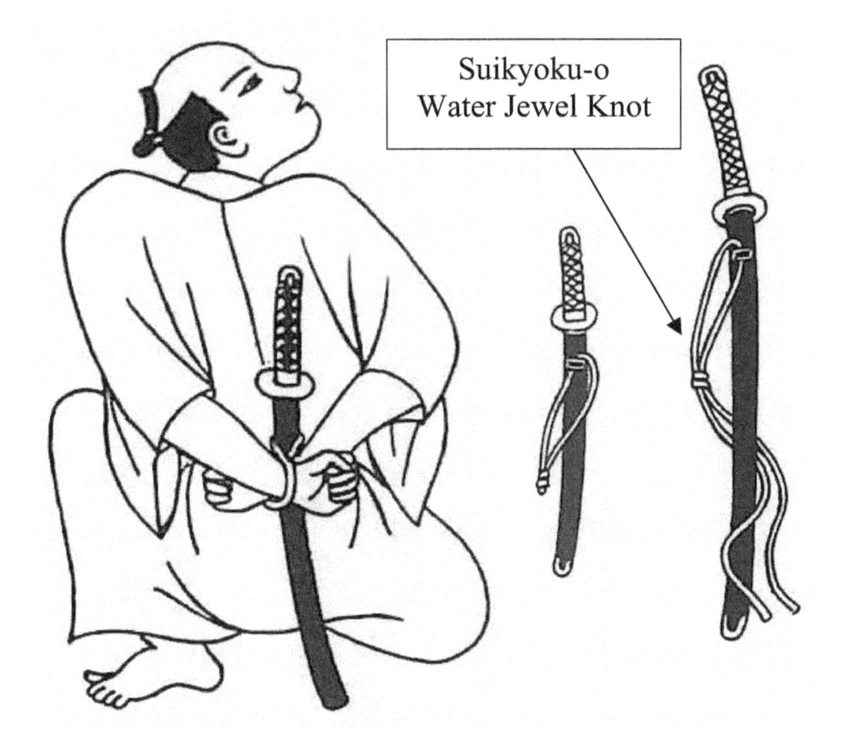
When you don't have any rope you can use the Sage-o cord from either the long sword or short sword. For a strong prisoner use this method. Insert both hands into the loop in the Sage-o known as Suikyoku-o, twist the rope and jam the Tsukagashira, pommel, into his back and this will prevent the prisoner from moving. If you raise his arms up you can then slide the Mu-Sori-Kaku-Tsuba, straight short sword with a square sword guard, into the loop formed by the hair. This will cause them to be immobilized. When moving the prisoner, jab the pommel into his back.

Ichiden School
Daisho Sageo Shibari #2 Using the Cord Attached to Either the Long or Short Sword

If you pull up on the Saya, the hands will raise up. You can then slide the sword into his hair.
Note: 無反角鍔 Mu-Sori-Kaku-Tsuba, straight short sword with a square sword guard.

Ichiden School

Torite no Koto
How to Capture

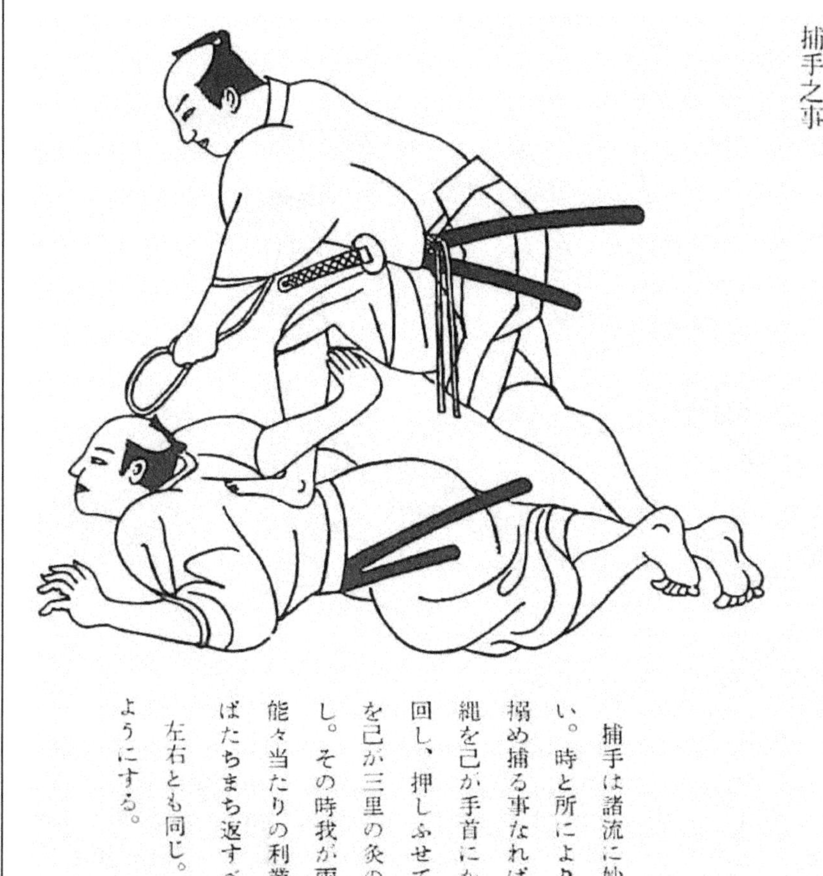

捕手は諸流に妙手あってもとより一定の法はない。時と所により、また、その人の技量によって搦め捕る事なれば、まず、当流にいうところの早縄を己が手首にかけ、敵の右手を取って、後背に回し、押しふせて敵の肩根を強く踏み、敵の手先を己が三里の灸の下通りに当たるごとく曲げてよし。その時我が両手あく故、縄さばき自由なり。能々当たりの利業には有之といえども、少し緩めばたちまち返すべし。必ず油断すべからず。

左右とも同じ。肩の踏どころ敵手を膈にて搦むようにする。

Ichiden School

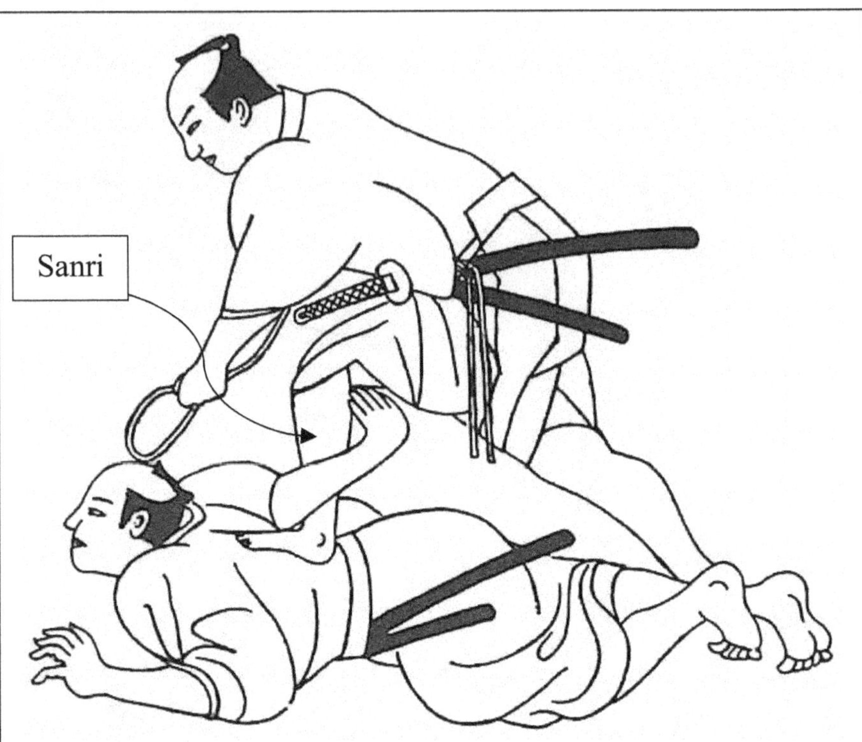

Torite, Capturing, is a highly technical method that is part of every martial arts school curriculum. However, the application varies by school. Depending on the time and place, as well as the skill of the person, the wrapping and tying will method will differ. First of all, in this school, wrap the Haya Nawa around your own wrist. Move around behind the prisoner and push him onto the ground. Step firmly onto the back of his shoulder joint. Pull his arm up and bend it over the front of your shin just below the acupuncture point known as Sanri. By doing this you will free up both hands to tie the rope. It is important that you position your foot and the opponent's arm carefully, if the pressure on the shoulder is weak or the elbow not firmly locked then the prisoner can roll over. It is essential you don't underestimate you opponent.

The technique is the same on the left and right. Use your shin to lock the prisoner's arm when standing on his shoulder.

Ichiden School

阿弥陀之胸割縄

当然極意の縛縄なり。

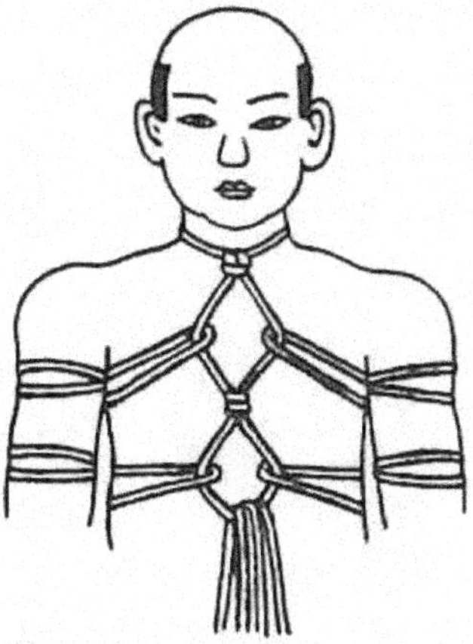

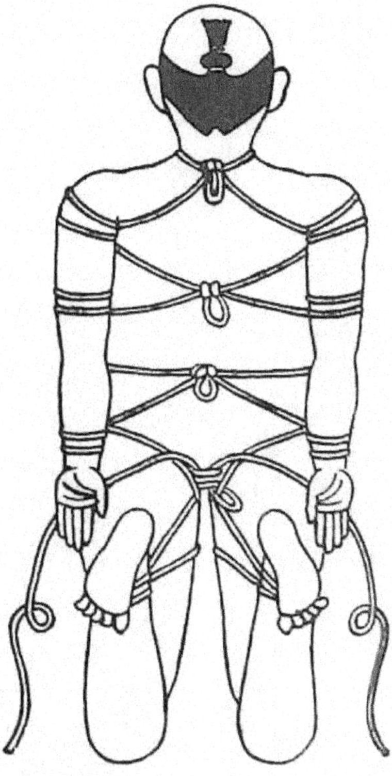

徴塵大極意

徴塵大極意というは当流捕手の極意なり。その場に有り合せたる何にてもとって敵に打ち付くべきなり。就中、灰などの類、火鉢、茶腕、火入れにても、眉間に投げ付けるなり。余は倣之。

<table>
<tr><td colspan="2">Ichiden School</td></tr>
<tr><td>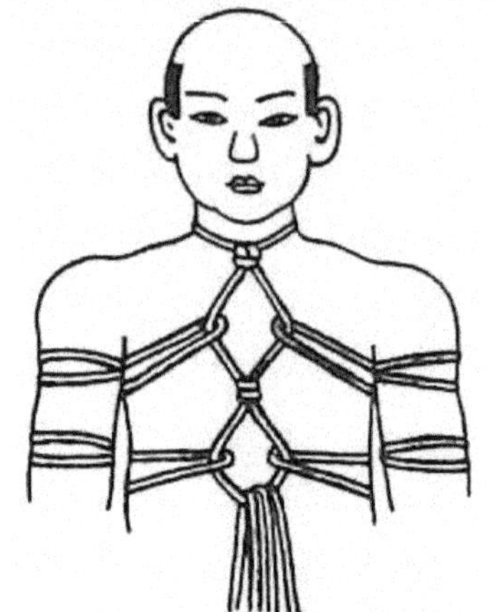</td><td>Amida no Mune Wari:
Split Chest Amida Buddha Tie

This is an inner mystery method of the Ichiden School</td></tr>
<tr><td>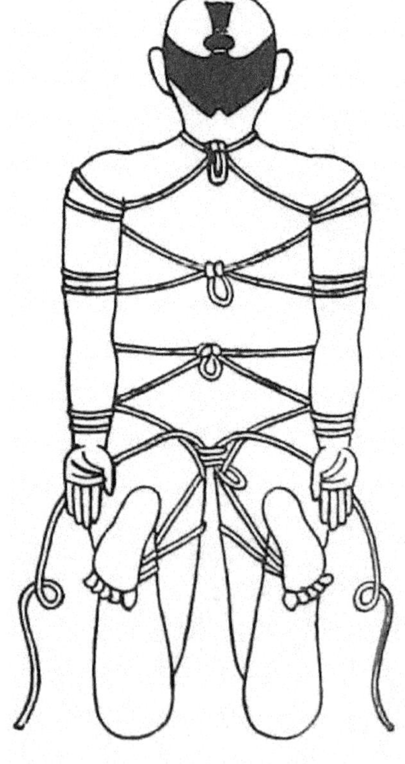</td><td>Mijin Dai Gokui:
Great Inner Mystery Pulverizer Tie

This is a great inner mystery technique of the Ichiden School. This method of arresting is a deep secret. Once tied you can do anything and hit the enemy with various objects. During this time you can throw ash, Hibachi brazier, teacups and so forth right between the eyes. The rest is up to you.</td></tr>
<tr><td colspan="2">An Illustrated Guide to Samurai Bondage Book One
End</td></tr>
</table>

Note:
A Brief Introduction to Lord Fudo

Lord Fudo is a protective deity in Buddhism whose name means "immovable protector." As Buddhism spread through China his name was transcribed with Kanji 不動 meaning "immovable." Lord Fudo was introduced to Japan by the Priest Kukai (774–835) also known as Kobo Daishi 弘法大師 *The Grand Master Who Propagated the Buddhist Teaching.* Miyeko Murase states in his 1975 book *Japanese Arts* that the origins of this Buddhist deity are in the Hindu god Shiva, particularly his attributes of destruction and reincarnation. In English his name is also written as King of Light, King of Luminescent Wisdom and Mantra King.

Lord Fudo is revered all over Japan and there are many temples dedicated to this deity. There are also over 130 waterfalls dedicated to Fudo around Japan. Lord Fudo is an important deity to ascetic monks like Yamabushi and practitioners of Shugendo and therefore many waterfalls are dedicated to him. Fudo is also important to Japan's Shingon Sect of Esoteric Buddhism.

Fudo is often shown with two attendants named Kongara Dōji 矜羯羅童子 and Seitaka Dōji 制多迦童子. The word Dōji means child. There are 19 Signs of Lord Fudo 不動十九観, a list that dates from the early 9th century.

	Japanese	Translation and Meaning
1.	大日如来の化身であること	He is an incarnation of Dainichi Nyorai. (He has a fearsome aspect to spur on those who resist enlightenment.)
2.	真言中に「ア」・「ロ」・「カン」・「マン」の四字があること	His Mantra has the four letters : a ・ ro ・ kan ・man
3.	常に火生三昧に住していること	He lives completely engulfed in flame. (The flame actually emits from his body burning away doubt and lighting the fire of wisdom)

4.	童子の姿を現わし、その身容が卑しく肥満であること	He appears youthful but fat and rather unpleasant. (Lord Fudo originated in India thus it reflects the ideal image of a youth, i.e. fat and healthy.)
5.	髪の毛の上に七沙髻があること	He has seven knots in his hair topped with a lotus. (The seven knots represent the Seven Paths to Awakening 七覚支 Mindfulness, Investigation, Energy, Joy, Tranquility, Concentration & Equanimity.)
6.	左に一弁髪を垂らすこと	His hair is in a bundle hanging over his left shoulder. (Shows compassion)
7.	額に水波のようなしわがあること	He has waves of wrinkles on his forehead (This was to show how he understood the suffering of the slaves in India, but this was changed to how he worries about souls in the afterlife in Japan.)
8.	左の目を閉じ右の目を開くこと	His left eye is closed, his right eye is open. (The left eye is closed because the "left road" leads to evil and Lord Fudo is showing that this way is blocked.
9.	下の歯で右上の唇を噛み、左下の唇の外へ出すこと	His lower tooth sticks into the upper right side of his lip and his lower left lip protrudes. (He does this to strike fear into the evil that tries to waylay humans)
10.	口を固く閉じること	His mouth is kept firmly shut. (Lord Fudo is showing the ideal manner, eliminating what is unnecessary and showing the ideal meditation pose.)

11.	右手に剣をとること	He has a sword in his right hand. (The sword cuts through the Three Poisons, sloth, rage and fear.)
12.	左手に羂索を持つこと	He has a rope in his left hand. (Originally a weapon for catching small birds or animals, it captures unseen evils.)
13.	行者の残食を食べること	He eats the leftover food of ascetic monks. (Encouraging people to avoid waste.)
14.	大磐石の上に安座すること	He is shown on a great stone slab. (This means one should be ever ready to receive enlightenment.)
15.	色が醜く、青黒であること	His skin is an unpleasant dark blue-green. (Japan shows Lord Fudo in red, blue and yellow as well. The colors are perhaps showing and exorcism.)
16.	奮迅して忿怒であること	He appears fierce and threatening. (He is showing intensity in contrast to the compassion of Kannon statues.)
17.	光背に迦楼羅炎があること	He has a flaming phoenix on as his halo. (This bird, which is ridden by Vishnu in the Hindu religion, eats the serpents that sew doubt.)
18.	倶利伽羅龍が剣にまとわりついていること	He can appear as a dragon wrapped around a sword. This is called Kurikara Ryu. (The earliest images of Lord Fudo were of a sword out of its scabbard.)
19.	２童子が侍していること	He has two child acolytes by his side. (Konkara Doji and Seitaka Doji, however an additional 36 children and 48 attendants are also shown.)

Tools of Lord Fudo 1:羂索 Kensaku

The coil of rope carried by Lord Fudo is called Kensaku 羂索 or Kenjaku. The purpose of it is to bind up the wicked or keep people from straying from the Buddha's teachings. A rope, made from five different colored strands (blue, yellow, red, black and white.) It often has a metal ring at one end, and one or more prongs, at the other.

Tools of Lord Fudo 2:降魔利剣 Gooma Riken

With this sword of wisdom, Fudoo cuts through deluded and ignorant minds and can subdue demons. The sword is also known as Sanko Ken 三鈷剣 which is a sword with a three-pronged Vajra (image below) as the hilt.

Other manifestations of Lord Fudo
倶利伽羅 Kurikara

Kurikara is a manifestation of Lord Fudo as a dragon eating a flaming sword. The Kanji refer to a "black dragon" but are written in the following ways:

倶利加羅
倶梨迦羅
古力迦羅
倶力迦羅

With this sword of wisdom, Lord Fudo cuts through deluded and ignorant minds and with the rope he binds those who are ruled by their violent passions and emotions.

Legend has it that Lord Fudo, as the representative of Buddhism, got into an argument with the representatives of 95 different religion. He changed himself into a flaming sword but one of the opponents did the same and the fighting went on without a winner. Then Lord Fudo changed his aspect into a fearsome Dragon Kurikara. Winding himself around the sword and grabbing with his feet and hands he began to eat the flaming sword, utterly humiliating his opponent.
Source: *The People's Guide to The Buddha* 庶民のほとけ
By Yoritomi Motohiro 頼富本宏著ＮＨＫブックス 1984

Illustrated Guide to Samurai Bondage

Book One

End

www.ingramcontent.com/pod-product-compliance
Lightning Source LLC
Chambersburg PA
CBHW031148270326
41931CB00006B/190

9781950959112